1,001
IDEAS FOR SCIENCE PROJECTS

1,001 IDEAS FOR SCIENCE PROJECTS

MARION A. BRISK, PH.D.
Center for Biomedical Education
at City College of New York

PRENTICE HALL
New York London Toronto Sydney Singapore

Prentice Hall General Reference
15 Columbus Circle
New York, NY 10023

An Arco Book

Manufactured in the United States of America

1 2 3 4 5 6 7 8 9 10

Library of Congress Cataloging-in-Publication Data

Brisk, Marion, A.
1001 ideas for science projects / Marion A. Brisk.
p. cm.
ISBN 0-13-633769-4
1. Science projects. I. Title. II. Title: One thousand and one ideas for science projects.
Q182.3.B75 1992
507.8—dc20 91-42529
CIP

CONTENTS

PART I

Helpful Hints

PART II

Ideas for Science Projects, Term Papers, and Reports

PART III

Resources

1,001 IDEAS FOR SCIENCE PROJECTS

TO THE READER

This book has two purposes: First, to provide you with a variety of stimulating, original ideas for science projects, research papers, and reports; second, to offer suggestions that will help you complete your particular assignment successfully. Many of the ideas presented here concern topics which are being discussed and studied both in the scientific and the lay community today. Some of these topics come straight from the headlines of recent newspaper and magazine articles. With so many topics to choose from, there are sure to be several that spark your imagination and interest.

The first part of this book, "Helpful Hints," contains suggestions on various aspects of your assignment. "How to Select Your Topic" describes the questions you should ask yourself in order to select a topic that you will both enjoy and also be able to complete successfully. "How to Research Your Topic" tells how to use a library and other sources effectively to obtain the information needed for your report or project. "How to Write a Research Paper" provides suggestions for writing a research paper and reporting the results of a science project following the scientific method which is critical to all scientific research. It includes general guidelines for scientific writing as well as tips for conducting scientific studies.

The second part of this book, "Ideas for Science Projects, Papers, and Reports," presents hundreds of exciting, timely project ideas. These ideas are categorized by scientific discipline—from Anthropology to Medicine to Space Sciences. Twelve scientific fields are represented in all, so you should be able to find a topic that intrigues you. Each topic is described briefly so that you can determine whether it is appropriate for your assignment and whether you find it interesting. A system of symbols is described in the first section. These symbols are used to help you identify various aspects of a topic that you should know about before you begin—the level of difficulty, for example, or the need for special equipment or supplies. To get you started you are provided with one or more journal articles, agencies, organizations or books that will be helpful in researching the topic you have selected. Remember, whether your assignment is to write a term paper or report or to do a project, you must first consult the scientific literature for what is known about the topic. All scientific studies begin with a literature search. The last part of the book, "Resources," lists the names and addresses of the scientific agencies or organizations that are mentioned in the ideas section. Often these agencies or organizations can provide you with written material including recent data that will be useful for your project or report.

PART I

Helpful Hints

HOW TO SELECT YOUR TOPIC

Selecting a topic is a critical step that should be done carefully. Here are suggestions for selecting an idea that you will be able to use successfully.

1. Select an idea that interests you

If you are particularly fond of biology, for example, you might first consider ideas in that field. Think about what particular topics interest you. Plant physiology? Microbes? Do you enjoy field work? Using a microscope? You might think back to when you last read a biology textbook and try to remember the topic or topics that you understood and enjoyed most. If you enjoy the topic you have selected, you are more likely to remain motivated as you research the scientific literature and ponder the details of your report or project. It is also a good idea to select a topic that is likely to be of interest to your reader or audience. Many of the ideas in this book relate to contemporary scientific issues that are of general interest.

2. Select an idea that you can complete in the allotted time

A project or report due at the end of a year will be more comprehensive than one due at the end of the semester. Therefore you must be sure that you will be able to complete your task in the allotted time and that you will have enough material to meet the requirements of your assignment. You may find, however, that you can be more specific about your assignment only after doing some initial reading from a general source like a textbook or review article. Try to be as clear and precise as you can about the title of your project or paper. You may find that you need to amend the exact title of your report or project as you begin to do the actual work.

Another important consideration is whether you can work on your assignment every day or only one or two days a week. Some projects require large blocks of time which you may not have available. Think carefully about the time demands of your project and make sure that they are compatible with your schedule.

3. Determine what special equipment or facilities will be necessary

If you are writing, for example, a very technical research paper make sure you have access to a library with extensive resources. A small community library most likely will not have the scientific journals and books you need. If your project requires special equipment or facilities investigate before you start whether they are available to you or can be made available to you. Your teacher or instructors may be able to provide you with the instruments or facilities you need. Be careful about any equipment or supplies you use in your home especially if young children or animals are part of your family. Before using any chemical look up its chemical properties and learn how it can be safely handled. This information can be found in the Merck Index which is usually available in university libraries as well as in other libraries.

4. Consider the level of difficulty of your topic

Make sure that you have a sufficient scientific background to write your paper or complete your project. If your topic is too technical then you may end up spending too much time gaining background information only to find that you do not have sufficient time left to complete your work. If you are unsure about your topic start to research it briefly before you choose it. It will become apparent that your topic is not suitable when you are unable to understand much of the material that you are reading. If this occurs do not hesitate to select a completely different topic or perhaps to make your topic more general so that you can avoid some of the more technical aspects of the subject.

To help select an appropriate project, a set of symbols appears next to each suggested topic. These symbols are listed below along with their meanings.

- ● This topic is of average difficulty and can be used for most high school and undergraduate assignments.
- Δ This subject encompasses extensive information and is only suitable for a long-term assignment.
- # This project requires special facilities, equipment, or supplies.
- ○ A large public or college library with extensive resources will be needed to adequately research this topic.
- * This project involves techniques and or supplies that could be harmful. Make sure you have adequate direction on how to safely use supplies and or perform procedures. Science instructors can often provide you with needed information.
- + This topic is very technical and will require previous knowledge.

HOW TO RESEARCH YOUR TOPIC

Once you have selected an idea begin by familiarizing yourself with your topic through general reference sources such as handbooks, dictionaries, and encyclopedias. Some general sources include the *Merck Manual*, *Van Nostrand's Scientific Encyclopedia*, and the *McGraw-Hill Encyclopedia of Science and Technology*. You can locate these sources in most libraries through the library card or microfiche catalogs. As you read from one of these sources outline the major concepts and be sure to indicate the source for future reference if needed. Index cards are especially useful for recording information from journal articles so that you can readily identify the source of your information when you list your references in your report.

After consulting a general reference locate a review article in one of the indices to scientific literature such as *Index Medicus* (medical or health related), *General Science Index*, or *Biological and Agricultural Index*. Review articles generally summarize controversies as well as accepted scientific principles involving your topic. They often will acquaint you with the terms and concepts that you need to understand about the scientific area you have selected. Review articles also cite important journal articles on your topic.

After you have read relevant articles go back to the subject headings in the indices to scientific literature, this time looking for more specific information. Use "see" and "see also" references for related materials. Some of these references you may already have obtained from the review article. If you are not sure of the titles of journals listed in the citations of the indices look up the journal abbreviations which usually appear in the front of an index. Copy citations accurately so that you can readily locate them later.

In addition to using the library to research your topic consult relevant organizations and agencies. Some of these are listed in the back of this book. Frequently, they publish recent information which may be available free of charge. You might even try writing to professors and scientists who lead laboratories that are conducting research on your topic. They may be of help in offering suggestions about your project.

HOW TO WRITE A RESEARCH PAPER

If you are doing a science project, you most likely will be required to submit a paper describing your work; so this chapter is important for you. If you are writing a term paper or report on a scientific topic this chapter is essential.

Scientific papers and projects generally follow the scientific method in their composition. Therefore writing a scientific report should follow the outline given below:

1. State Your Objectives

Begin your paper by clearly indicating the goal or goals of your report or project. Writing this introductory section early on is very important because it will help you crystallize in your mind what your report or project will investigate. It is a good idea to refer to this section frequently as you work on your paper or project to insure that you remain on course. Usually background information along with references are included in this introductory section to support the validity of your paper or project.

2. Describe Your Procedure

It is a good idea to plan your procedure or method. If you did your research in libraries, what special problems did you encounter? If you did a field or laboratory study, describe all techniques and methods used. Outline proposed techniques and methods, and make appropriate changes as your project progresses. Most likely you will find that you will digress somewhat from your original plan as your project develops because you will find more effective techniques or perhaps you will encounter unexpected problems. If you used interviews or questionnaires to gather information be sure to include the details of their administration. (A copy of questionnaires and interviews should be appended to the paper.)

3. Report Your Results

Following a discussion of your procedure, results are usually given. If possible use graphs or tables to summarize your results. Be sure to include only what you actually found, not what you would like to find or thought you found. Carefully label all graphs and tables so that the meaning of all given values is clear. It is often helpful to think about how you might display the data even before you complete the project so that you are clear in your own mind about what you are attempting to determine.

4. Discuss Your Results

This section is critical. Here is where you show how your results have met your objectives. What are your conclusions? Remember these must be based on your data. What is the significance of your findings and conclusions? What new problems, if any, are raised by your study that suggest new areas of research? Do your results confirm previous studies or are there discrepancies? Be specific and carefully support all statements.

Some General Guidelines for Scientific Writing

1. Be Clear and Direct

State facts and ideas in a clear and direct manner. Avoid use of jargon or complicated sentences. If you use technical terms, make sure they are explained thoroughly.

2. Do Not Assume Your Reader Knows Your Topic

State and explain your ideas clearly. Do not assume that your reader knows what you mean. Think of your reader as an intelligent person who has little knowledge of your subject. You are always better off providing too much information for your reader than too little so that he or she is able to understand your work.

3. Avoid Bias and Generalizations

Support your conclusions with data, not personal opinions. If you are conducting a study, let your data lead you to your results. If you offer an explanation for your results, try not to let your personal opinions influence your work. Avoid generalizations that encompass subject areas not included in your report or project.

4. Do Not Make Your Report Too Long

Students often assume that longer reports will produce higher grades. This is not true. The length of your paper should depend on your particular project. Do not include superfluous information just to add length. On the other hand, be sure to provide sufficient information to support your ideas, to present what is already known about your topic, and to explain your results and conclusions. Be sure to include all relevant information to fully support your statements. Again you are better off providing too much information than too little information as long as it is relevant.

5. Use Appropriate Documentation

Be careful to document the words and ideas of others. Do not represent the words or ideas of others as your own by failing to document appropriately. The following general guidelines will help you to know when to include references to the work of other authors.

— Always use your own words to express your ideas. Whenever you read an article or section of a book, take notes using your own words to summarize important information.

— Be sure to document the experimental results of others as well as their conclusions and explanations of their results. In general you do not need to document facts that are common knowledge.

— If you are going to use the words of another person in your paper make sure that the quoted, written or spoken words are reproduced exactly.

— Plagiarism means using the words or ideas of another as your own. This includes having your family or friends write your paper for you.

— The most common method for attributing work you describe to the appropriate authors is to use numerical superscripts following the relevant passages. These references are then listed on a separate sheet at the end of your paper. Each listing should include the author(s) names, journal or book, volume, year of publication, and page number from which the information was taken. The actual format of the references will depend on the requirements of your course.

6. Use Graphs, Charts, and Tables Effectively

Graphs, charts, and tables can be very useful in highlighting and clarifying results and data. They can increase the reader's understanding of your work because they often convey trends, comparisons, and relationships more effectively than written material. It is critical, however, to display your data using the appropriate graph, chart or table. If you are unsure about what kind of illustration to use, consult your instructor or the literature. Refer to the work of other researchers in your field and see how they presented their results. The following reference may also be helpful:

The Visual Display of Quantitative Information
Graphics Press, 1983
Cheshire, CT
E. Tufte

Tables are used when the data are precise numbers and in particular should be used when there are too many values to be included in the text. Tables can convey trends that are not easily recognized when data is presented as narrative. Tables should always be simple and as brief as possible. Titles and column headings should be concise and clear. Use abbreviations and symbols to keep headings short, but make sure to use the same abbreviations and symbols that appear in your report.

Tips on Conducting Scientific Studies

Below are four steps you should follow when conducting any kind of scientific investigation.

1. Know Important Properties of All Materials Used

Make sure you are aware of any harmful properties of any reagents or materials you use. Look them up in the Merck Index, which is generally kept in the reference section of many libraries. If any of the materials you are using can be harmful, seek appropriate supervision; do not use harmful chemicals in your home. Also, be sure to find out how to safely dispose of all materials after you are finished with your experiments.

2. Learn How To Use All Equipment and Instruments Properly

It is extremely important that you learn how to use all equipment and instruments necessary for your project before you begin to work. Improper use of equipment will not only adversely affect your results but also may be dangerous. In addition, keep in mind that scientific instruments are generally fragile so that improper useage may result in breakage and prevent you from completing your project. It is always a good idea to practice using all equipment before you begin your experiments.

3. Make Sure You Understand All Procedures Completely

Before you begin your study make sure you understand thoroughly all procedures to be used. This is as true for a chemical test as it is for conducting an interview for a survey. Go through several trial runs acquainting yourself with the procedure and identifying the more difficult aspects of the procedures. You may find that parts of your procedures need to be either completely or partially changed.

4. Decide How You Will Collect and Record Your Data

Before you begin your study determine how you will collect and record your data. If possible construct tables that you can just fill in as you conduct your investigation. Make sure that they are completely labeled so that you will never question the meaning of an entry. Never record your data on loose papers. Use the same notebook to record all information regarding your project.

PART II

Ideas for Science Projects, Term Papers, and Reports

1

ANTHROPOLOGY

EVOLUTION OF MAN • Δ

Various theories have been proposed about the evolution of modern humans. Some scientists believe that hominids (evolutionary family that includes modern humans) diverged from African great apes between 5 million and 10 million years ago. A team led by Louis de Bouis of the University of Poitiers in France, however, suggests that hominids diverged from African gorillas and chimpanzees about 12 million-years-ago. They base their theory on a 9–10 million-year-old fossil skull recently discovered in Greece which they propose came from a direct ancestor of hominids. They claim that the skull may be the earliest discovered example of a hominid. The size and shape of its teeth are suggestive of *Australopithecus afarensis*, the 3.5 million-year-old African hominid group in which the remains of "Lucy" are included. Several prominent anthropologists feel that the recently discovered fossil does not display clear anatomical links to hominids and so the debate about the evolution of humans continues.

IDEAS TO EXPLORE

- What are the various theories proposed to describe the evolution of man? What data are they based on?
- Who are the main players in this debate?
- Who is Richard Leakey and what is his theory?
- Who is "Lucy"?

Scientific Techniques of Anthropology

To date animal and human remains anthropologists often rely on techniques which measure concentrations of some radioactive isotopes.

IDEAS TO EXPLORE

- What are these isotopes? How can their measurement reveal the age of the remains?
- What other methods or techniques are used?
- What errors or uncertainties are intrinsic to these techniques?

SOURCES

Scientific American, Dec. 1990, p. 98.

SYMBOL KEY	
•	Topic of average difficulty
Δ	Long-term assignment
#	Project requires special facilities, equipment or supplies
○	Large public or college library required
*	Safety precautions required
+	Highly Technical; specialized knowledge required

2

ASTRONOMY

THE MODERN PICTURE OF THE UNIVERSE • Δ

At the time of Edwin Hubble (1920's) a fierce debate was raging amongst astronomers concerning spiral nebulae—luminous and usually hazy swirls that appeared scattered throughout the sky. One side believed that nebulae were formed within our galaxy and consisted of small clouds of gas and dust illuminated by interior stars. Other astronomers supported the view that these citings were really large rotating star systems like the Milky Way and were located far away from the boundaries of our galaxy. Hubble with his powerful telescope resolved the controversy by photographing Cepheids (pulsating stars) in the Andromeda nebula. These changing stars which expand and contract can be used to calculate the distance between Andromeda and us. Hubble showed that Andromeda was located far beyond the limits of our galaxy, proving that other galaxies exist in the universe. Hubble also showed by measuring the red shift of detectable galaxies that the universe is continually and uniformly expanding. Hubble's law states that the speed at which detectable galaxies are travelling away from the earth is directly proportional to their distance from the earth.

Today astronomers attribute this outward flow of galaxies to the Big Bang—the explosion that occurred some fifteen billion years ago which triggered the expansion of the universe. In addition, other motions of galaxies have been detected aside from those associated with the Big Bang. These independent motions have been attributed to strong gravitational forces. Such a powerful force is thought to be pulling on the Milky Way and has been called the Great Attractor.

IDEAS TO EXPLORE

- What is known about the Great Attractor?
- How are galaxies now known to be distributed in the universe?
- What is the dark-matter model which is used to explain the superclusters of galaxies in the universe?
- How does dark matter relate to the motions of spiral galaxies?

SOURCES

The Sciences, Sept./Oct. 1989.

A. Pressler
Observatories of the Carnegie Institution of Washington, in Pasadena, CA

SYMBOL KEY	
●	Topic of average difficulty
Δ	Long-term assignment
#	Project requires special facilities, equipment or supplies
○	Large public or college library required
∗	Safety precautions required
+	Highly Technical; specialized knowledge required

DO OTHER PLANETARY SYSTEMS EXIST? ☉

Astronomers have searched for other planetary systems in the solar system for decades. In 1916 Edward Barnard photographed a reddish star (Barnard's Star) whose unusual motion (called "wabbles" by astronomers) he proposed resulted from the gravitational force of two orbiting planets. More recently other astronomers (Bruce Campbell and David Latham, for example) noted fluctuations in the motion of stars which they attributed to unseen planets. Planets give off less reflected light than stars and are much less massive. Consequently, they are very difficult to observe given the limits of the most powerful telescopes available today.

IDEAS TO EXPLORE

- Does our present notion of how the solar system formed infer the existence of other planetary systems?
- What recent discoveries indicate the existence of planets orbiting stars other than our sun?
- What advances in instrumentation will be needed to detect other solar systems?

SOURCES

The Sciences,, May/June 1989, p. 31
D. Black—Lunar and Planetary Institute in Houston, Texas

E. Levy—Planetary Sciences at University of Arizona, Tucson, Arizona

D. Latham, Smithsonian Astrophysical Observatory in Cambridge, Mass.

D. Backman, Kitt Peak National Observatory, Tucson, Arizona

SYMBOL KEY	
●	Topic of average difficulty
Δ	Long-term assignment
#	Project requires special facilities, equipment or supplies
○	Large public or college library required
*	Safety precautions required
+	Highly Technical; specialized knowledge required

ENERGY OF THE SUN [+ ○]

The radiation released by the sun comes from the fusion reactions at its center. The fusion reactions that are thought to occur, however, according to calculations should release 2% of their energy in the form of neutrinos. The number of neutrinos actually detected on Earth is much lower than what is predicted. This deficit in the number of neutrinos detected makes suspect the whole theory of how stars generate energy. One explanation is that solar neutrinos may behave in unexpected ways and thus avoid detection.

There is much controversy over the properties of neutrinos including their mass and magnetic moment. Neutrinos are thought to exist in three forms or "flavors." Perhaps by a transformation or change of flavor some neutrinos are not detected if an instrument is designed to count only one type of neutrino. A proposed neutrino observatory in Sudbury, Ontario would be able to detect transformations of neutrinos.

Another explanation to account for the solar neutrino deficit calls into question the standard solar model which is used for calculations. Some "nonstandard" models are now being used in which the sun's composition or internal temperature among other variables differs from those assumed in the standard model.

IDEAS TO EXPLORE

- What fusion reactions are thought to occur in the center of the sun?
- Where do the neutrinos come from? What is known about neutrinos?
- What is the structure and composition of the sun and how do these characteristics relate to emission of neutrinos?
- What experiments are presently being conducted to resolve the case of the missing neutrinos?

SOURCES

Physical Review Letters, July 3, 1989

Physics Today, July 1989
E. W. Beir
University of Pennsylvania

SYMBOL KEY	
●	Topic of average difficulty
Δ	Long-term assignment
#	Project requires special facilities, equipment or supplies
o	Large public or college library required
*	Safety precautions required
+	Highly Technical; specialized knowledge required

SUNDIALS: TIME BY THE SUN ●○

There are an enormous number of different types of sundials. Many can be used to measure other than local mean solar time. Analemmatic and horizontal dials combined can be used to locate north and south while the Capuchin dial indicates the times of sunrise and sunset.

IDEAS TO EXPLORE

- What kinds of different dials are there and how do they each work? What can they be used for? When were they developed?
- How can sundials be used to measure the value of the equation of time on any one day?

PROJECT

Make Your Own Sundial

Construct one or more sundials. The simplest sundial is called the equatorial dial. It can be made by putting the dial in the plane of the celestial equator with a pointer (gnomon or style) in the center. To construct an equatorial sundial follow these directions:

(1) Draw a circle with a diameter of 15 cm on a plain piece of cardboard.

(2) Using a protractor, draw diameters across the circle at 15° intervals. Each diameter represents an hour line.

(3) Use the right-hand end of one diameter as the 6 hour line. Proceed in a clockwise direction, marking the next line as the 7 hour line and so forth until you have completed the circle with the 18 hour line. The lowest point on the dial should be the 12 hour line.

(4) Insert a long needle in the center of the circle to serve as the gnomon, or pointer, of the sundial. The pointer should be perpendicular to the face of the dial.

(5) Support your sundial by making a stand out of two pieces of cardboard. The angle of the face of the dial to the ground must be equal to that of the equator at your latitude. This angle is the co-latitude and is equal to 90° latitude.

(6) To set up your sundial rotate it until the shadow of the needle falls on the local solar time.

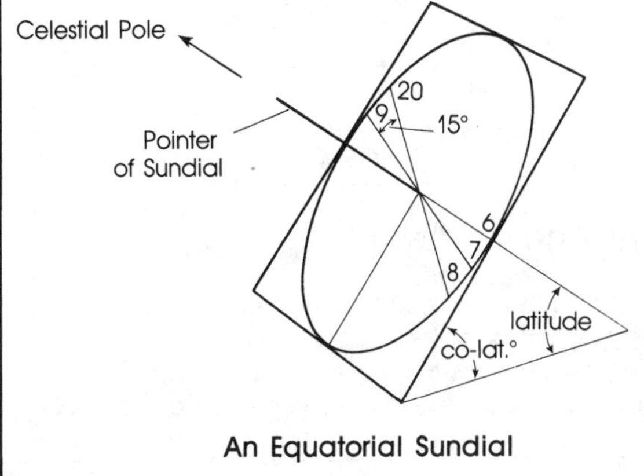

An Equatorial Sundial

You can use your sundial to perform a variety of experiments including using the moon's shadow to tell the time at night, locating due north and south, and determining sunrise and sunset times. For detailed directions on the construction and use of sundials consult:

Experiments in Astronomy for Amateurs,
Richard Knox
St. Martins Press, Inc., 1976.
New York, NY

SYMBOL KEY	
•	Topic of average difficulty
Δ	Long-term assignment
#	Project requires special facilities, equipment or supplies
o	Large public or college library required
*	Safety precautions required
+	Highly Technical; specialized knowledge required

MARS: THE SEARCH FOR LIFE IN THE SOLAR SYSTEM ⊡

Among the planets of the solar system, Mars has always been of special interest. Of all of the planets Mars has been suspected of harboring extraterrestrial life mostly because of its closeness to the earth and because its atmosphere, unlike that of Venus, is often clear revealing a surface that appears amazingly earthlike. Percival Lowell furthered the life on Mars hypothesis. He founded the Lowell Observatory in Flagstaff, Arizona, built for the purpose of studying Mars. Lowell claimed to have seen a complex system of artificial canals on the Martian surface and attributed the construction of these canals to highly intelligent beings. He estimated, based on his interpretation of his observations, a Martian atmospheric pressure of 85 millibars which might be sufficient to sustain some form of life. Although Lowell's views were clearly discredited, many astronomers continued to support some of his findings despite the undeniable evidence. Consequently, NASA (National Aeronautics and Space Administration) made the exploration of Mars a top priority in its planetary missions. The Lowellian dream of extraterrestrial life actually fueled the development of the space missions beginning with Mariner 4 and ending in 1976 with Viking 1 and Viking 2. The results of these missions have shown that Mars cannot presently support life although there is some evidence that it may have in the past. The seasonal ice caps that had been observed from earth turned out to be frozen carbon dioxide and the markings on the Martian surface attributed to organic matter are desert areas periodically covered with dust. The canals were formed by natural physical forces.

IDEAS TO EXPLORE

- What evidence did the early astronomers present to tantalize the imagination with the possible existence of highly intelligent Martians?
- What did the various spacecrafts visiting Mars reveal about the true nature of the planet's surface and atmosphere?
- The atmospheric pressure and temperature on Mars indicate that water must be frozen as a permafrost. Was Mars warmer in the past? Was its atmosphere denser? Could some form of life have existed?

PROJECT

The Topography of Mars ● ○

For more than two years landers delivered by the Viking spacecrafts have provided hundreds of pictures of the Martian surface. The southern hemisphere of Mars contains enormous mountains and craters including the volcanic mountain Olympus Mons which is three times the height of Mt. Everest. The northern surface of the hemisphere is relatively smooth.

Construct a projection of Mars (two dimensional map) showing the topography of each major region including names of major mountains, craters, canyons, etc. Use a system of symbols or colors to indicate size of these structures or formations. You can select a global projection of the earth to use as a model.

SOURCES

The Solar System
Scientific American Library
R. Smoluchoruski

Orbiting the Sun: Planets and Satellites of the Solar System
Harvard University Press
F. Whipple

The Sciences, March/April 1990, p. 45.
N. Horowitz

SYMBOL KEY	
●	Topic of average difficulty
Δ	Long-term assignment
#	Project requires special facilities, equipment or supplies
o	Large public or college library required
∗	Safety precautions required
+	Highly Technical; specialized knowledge required

3

BIOLOGY

CIRCADIAN RHYTHMS ⊡

Circadian rhythms or internal biological clocks have been studied in many species. For example, it has been shown that if humans are not given any hint as to what time of day it is our sleep/wake pattern, and physiological functions all follow a rhythm of approximately 25 hours. Czeisler and Krondier from Harvard Medical School showed that our internal clock is sensitive to light so that ordinary room light can be used to synchronize the human circadian system to a 24-hour day. More recent work implies that bright light could be helpful in correcting sleep disorders as well as promoting recovery from jet lag. Dr. Czeisler reset the biological clock of the crew members of the space shuttle *Columbia* before their scheduled launch so that they could work the night shift.

IDEAS TO EXPLORE

- What is known about biological clocks in different species?
- How are biological clocks related to physiological functions?
- Do plant species exhibit internal clocks?
- What is known about the human circadian rhythm? What physiological functions are affected?
- How has light been used to correct internal clocks? How are psychiatrists using phototherapy to help some patients?

PROJECT

Effect of Light on Plant Growth and Development ⊡

Light has been shown to change the internal clock of humans so that night-time hormonal levels and other physiological functions are at morning levels. Use an analogous protocol to observe the effects of light on plant growth and development. Select a fast growing house-plant, for example, and use a dark room or cardboard boxes to cover the plants. Expose several plants to light following the existing day/night cycle. These plants will serve as the controls. Expose other plants to light only at night, reversing the usual day/night cycle. Compare the growth and development of these two groups of plants.

SOURCES

Science, June 1989, p. 1256

Nature, March 1991
C. Czeisler, et al.
Brigham and Women's Hospital at Harvard

Science, vol 244, 1989, p. 1328
C. Czeisler, et al.

HOW ENVIRONMENTAL STIMULI EFFECT PLANT GROWTH ⊡

Plants are known to react to environmental pressures such as wind, rain, and even human touch. Coastal trees, for example, become shorter and stronger in response to strong winds and heavy rainfall.

In a laboratory study at Stanford University the same changes in growth patterns were induced by touching plants twice daily. They also found that the mechanism for these growth changes involved an increase in messenger RNA (mRNA). This gene activation did not occur unless there was direct stimulation.

IDEAS TO EXPLORE

- What is known about the effect of direct stimulation on fruit production of plants?
- What are the effects, if any, of radiation, electromagnetic fields or sound on plant growth?
- What biochemical or physiological mechanisms have been suggested to account for a plant's response to stimulation?

PROJECTS

1. Effect of Stimuli on Plant Development and Fruit Production ⊡

Determine the effect of stimuli on plant development and on production of fruit. Choose a plant that can be grown indoors so that you can control stimulation from the environment. You might attempt to measure the effect of human touch on the production of fruit or on growth of leaves and stems in bean or tomato plants. A fan or hair dryer can be used to imitate wind. Does watering plants overhead versus watering only the surrounding soil affect growth and production?

2. Effect of Electromagnetic Fields on Fruit Production or Growth • Δ

For a more comprehensive project another group of plants could be used to determine the effects of electromagnetic fields on fruit production or growth. Copper wire surrounding test plants and connected to a battery will establish an electromagnetic field. When testing the effects of one, some, or all of the above mentioned stimuli be sure to have a control group which is receiving the same nourishment, light, and water.

SOURCES

Cell, Feb. 9, 1990.
R. Davis and J. Broam
Stanford University, CA.

WAVELENGTHS OF LIGHT AND SEED GERMINATION • Δ # *

The wavelengths of incoming solar radiation range from 200 to 3,000 nanometers (one nanometer is a billionth of a meter). Ultraviolet radiation is between about 3 and 380 nanometers (nm) while the visible region extends from about 400 nm to 700 nm with violet at the lower end (greater energy) and red at the greater wavelength region (lower energy).

To see the effect of light of different wavelengths on seed germination you could do one of the following projects.

PROJECTS

1. Effect of Blue Light *vs.* Red Light

• Δ # *

Select seeds from a plant such as tomato, cucumber, or bean and expose several batches of seeds to ultraviolet light, several to blue light and several to red light.

2. Effect of Ultraviolet Radiation on Different Plants • Δ # *

Irradiate several batches of seeds of several different plants (tomato, cucumber, bean, squash, etc.) with ultraviolet radiation and then plant. Because of increasing UV radiation reaching the earth due to depletion of the ozone layer the effect of UV light on plant life is of concern to us all. Your project may show that seeds of one plant are more susceptible to UV light than others. (If you are interested in flowers use marigold or other seeds which can be started indoors.)

3. Effect of Ultraviolet Light on Plant Growth • Δ # *

Select seeds of one plant and irradiate several batches with UV light. Measure not only rate of germination but also monitor the plant growth and fruit production, if possible, relative to control groups.

For all of the above mentioned research projects the period of exposure necessary to produce a measurable effect (if any) is unknown. Therefore you may need to increase exposure periods and repeat the experiment. Make sure you use control groups and try to keep important variables such as water, soil, exposure to sunlight, and humidity constant for all experimental batches and their control groups. Also make sure you provide the appropriate growing environment for the seeds. To increase the number of seeds you irradiate and plant you could irradiate the seeds and distribute them to friends and family or other interested participants along with control seeds to plant.

SOURCES

Scientific American, Jan. 1989, p. 88.
P. Moses and N. Chua

WAVELENGTHS OF LIGHT AND PLANT GROWTH ⊡

Visible light consists of many wavelengths each having its characteristic color. It has been recently shown that exposure of tomato plants to red light induces the growth of small tumors on their leaves.

PROJECT

Select a different plant such as a bean plant or even a fast-growing houseplant and expose groups of plants to a particular light. Measure the rate of growth of each plant as well as leaf and fruit production. Compare with a group of control plants exposed to white light. The different colored lights can be generated from filtered lamps.

SOURCES

Science News, Jan. 14, 1989, p. 23

Plant Physiology, Dec. 1989
T. Tibbitts and R. Morrow

SYMBOL KEY	
•	Topic of average difficulty
Δ	Long-term assignment
#	Project requires special facilities, equipment or supplies
o	Large public or college library required
*	Safety precautions required
+	Highly Technical; specialized knowledge required

ALTERNATE METHODS OF PEST CONTROL ⊡

Since the "green revolution" of the 1960's farmers have used chemical pesticides and herbicides to control unwanted insects and plants in order to increase their yield of desired crops. However, many of these largely organic chemicals have had deleterious effects on the environment, polluting our drinking water and destroying fish and wildlife. As a result, alternate pest control measures are being developed which either decrease the amount of chemical needed (integrated pest management) or replace the harmful pesticide. Biological pest control techniques include introducing a predator, developing plant species that are resistant to attacks by the pest, and sterilizing insects in captivity then releasing them into the environment to mate with wild forms.

IDEAS TO EXPLORE

- What are the various classes of pesticides? What are their general structures and what is known about the mechanisms associated with their efficacy?
- What are some known environmental impacts of these substances? What alternate control measures have been introduced and are also now being developed? Give specific examples.

PROJECTS

1. Alternate Methods of Pest Control for a Farm [•]

Design a farm that incorporates alternate methods of pest control. Give details of the farm's operation throughout the year.

2. Improving Pest Control Strategies [•]

Select a farm in operation today not using integrated pest management and design improved strategies for pest control. Prepare an environmental impact report on the farm.

3. Pest Control in Suburban or Urban Areas [•]

If you live in a suburban or urban area where residents have gardens, assess the impact of their methods of pest control on the environment, e.g., groundwater pollution, urban runoff. What alternate programs could they use to minimize their impact?

SOURCES

Environmental Science
J. Turk and A. Turk
Saunders College Publishing, U.S.A.

COMPOSTING •

Composting on both an individual and municipal level is an important aspect of recycling and decreasing pollution. Composting not only decreases landfill useage but provides a natural source of fertilizer which will not pollute water and deplete the soil. Inorganic fertilizers leach out of the soil, entering surface and groundwaters. Dangerous nitrate levels often appear in drinking water in farm areas. Inorganic fertilizers also decrease the fertility of agricultural lands.

Composting experiments are needed to determine the effects of temperature, moisture, oxygen exposure, and organic content on the rate of decomposition and quality of the end product.

PROJECT

Creating Organic Fertilizers •

Create small compost piles in large glass vessels (beakers or flasks). Use different combinations of shredded kitchen and backyard wastes (leaves, sawdust, etc.) to vary organic content. Also vary temperature and moisture of the piles. Then test the content of the end product of the compost materials with a test-kit that you can usually obtain from any nursery or garden store. These test kits are used to determine the fertility of soils. They include measurement of pH, nitrogen and potassium concentrations.

SOURCES

There are many books available on composting in local nurseries, bookstores, and libraries. Also many articles have appeared in *Organic Gardening*. Write to the U.S. Department of Agriculture or your state department of agriculture. Organizations which promote sustainable agriculture will also provide you with information. For example,

The International Alliance for Sustainable Agriculture
1701 University Avenue SE,
Minneapolis, MN 55414
(612) 331-1099

SYMBOL KEY	
●	Topic of average difficulty
Δ	Long-term assignment
#	Project requires special facilities, equipment or supplies
o	Large public or college library required
∗	Safety precautions required
+	Highly Technical; specialized knowledge required

THEORIES OF EVOLUTION: WAS DARWIN RIGHT? • Δ ○

A basic tenet of neo-Darwinian theory is that all mutations arise spontaneously and randomly independent of environmental conditions. Recent evidence, however, suggests that bacteria may develop more mutations that help them survive than those that do not. Such a "directed" mutation theory poses a challenge to classical evolutionary theory and consequently has stirred a vigorous debate amongst evolutionary biologists.

IDEAS TO EXPLORE

- What is Darwin's theory of evolution? What recent research challenges his theory?
- What other explanations have been offered to explain this recent data about the nature of mutations? What additional experiments can be conducted to test this new theory?
- What other challenges to evolutionary theory have appeared in the literature? What evidence is there to support these theories?

SOURCES

Genetics, Sept. 1990
B. Hall
University of Rochester

THE EFFECT OF REDUCED SIZE ON VERTEBRATE SURVIVAL ⊡

The idea that extreme size, large or small, hampers survival has been a popular theme of evolutionary biology. For example, it is thought that the excessive energy requirements and small brain of the huge dinosaur *atlantosaurus* led to extinction. Decreased body size is also thought to be detrimental to a species' survival. At first miniaturization of a species was taken to be indicative of decline of that species. Recent studies however have indicated that miniaturization can enhance survival. The Mexican salamander of the genus Thorius is an example of longterm survival despite the fact that its body may be the smallest possible for a vertebrate.

IDEAS TO EXPLORE

- How can the size of an organism be markedly reduced without affecting the organism's ability to function?
- What structures must be redesigned or eliminated to accommodate reduced size?
- How can reduced size be the key to survival of some species?

SOURCES

The Sciences, Sept./Oct. 1986, p. 43.
J. Hanken

Science News, *136*, 357 (1990).
A. Lister

EXTINCTION: THE COLLAPSE OF BIOLOGICAL DIVERSITY • Δ

Today animal and plant life are being destroyed at an unprecedented rate. Endangered species range from the tiny *tree frog* to the African elephant and rhino. These species may face extinction because of destruction of their habitat, human predation, or pollution. For example, the African and Asian rhinos are slaughtered for their horns which are used for traditional medicine in some Asian cultures. Black rhinos in particular have suffered a severe decline in population from 65,000 individuals to only about 4500 today, threatening the species with extinction.

IDEAS TO EXPLORE

- In what regions of the world are species disappearing at an alarming rate? Why?
- What are the consequences of such a devastation of biological diversity? What is being done to preserve the species?

SYMBOL KEY	
•	Topic of average difficulty
Δ	Long-term assignment
#	Project requires special facilities, equipment or supplies
o	Large public or college library required
*	Safety precautions required
+	Highly Technical; specialized knowledge required

PROJECT

Examine the Decline of a Species [•]

Select a species that has become extinct or is threatened with extinction and research the reasons for its decline. Predict the effects of the decline on the ecosystem. What efforts are being made to stop the decline? Can you offer some solutions? Give examples of species that were saved from extinction.

SOURCES

Science, Oct. 10, 1986, p. 147.

U.S. Department of Interior, Fish and Wildlife Service
Washington DC

World Wildlife Fund
1250 24th St NW
Washington DC 20037

GLOBAL DECLINE IN AMPHIBIAN POPULATION ⊡

For the last 10 to 20 years some species of frogs, toads, and salamanders have been dying in large numbers. This documented decline in amphibian populations has been attributed to acid rain and snow, global climate changes, chemical pollution, and stocking lakes with game fish who are natural predators of tadpoles. It has been suggested that the decline of amphibians is indicative of a global problem which will affect the survival of other species including man.

IDEAS TO EXPLORE

- Which species of amphibians have shown a rapid decline? Describe their habitats as well as possible causes of their demise.
- Why might amphibians be particularly sensitive to environmental stresses?

SYMBOL KEY	
•	Topic of average difficulty
Δ	Long-term assignment
#	Project requires special facilities, equipment or supplies
○	Large public or college library required
∗	Safety precautions required
+	Highly Technical; specialized knowledge required

PROJECT

Local Amphibian Behavior

Locate a pond or lake in your community or in a public park nearby. Observe amphibian activities and discuss possible population changes over the last years with park rangers, local biologists, or naturalists. Determine if conditions exist which could lead to a decline such as a low pH, chemical pollution (urban runoff, etc.) or increase in predators.

SOURCES

Science News, Feb. 24, 1990, p. 116.

THE GAIA HYPOTHESIS ☉

The Gaia hypothesis, named after the Greek goddess of earth, was proposed by James E. Lovelock of the Coombe Mill Experimental Station in Cornwall and Lynn Margolis of Boston University. The hypothesis attributes the earth's climate to the role of biota which exchange carbon dioxide with the atmosphere. It accounts for the decrease in atmospheric carbon dioxide as a result of living organisms.

IDEAS TO EXPLORE

- What are the main ideas of the Gaia hypothesis?
- How does the Gaia hypothesis explain the evolution of the earth's atmosphere? What data is used to support this hypothesis?
- What is the role of the carbonate-silicate cycle in modulating the earth's climate?
- According to the Gaia hypothesis what may be the effect of the unprecedented decline in animal and plant life on earth due to human activity?

SOURCES

Gaia: A New Look at Life on Earth, 1979
Oxford University Press, New York
J. E. Lovelock

THE LOSS OF BIODIVERSITY AND ITS CONSEQUENCES ⊡

Five mass-extinction events are known to have occurred since the advent of life on earth. Our species evolved during a period of the greatest biological diversity (biodiversity) in the history of the earth. Today, however, biodiversity is shrinking at an unprecedented rate as a consequence of human activity.

It has been estimated that we have studied about 1.4 million species, but that as many as 30 million species may exist worldwide. More than half of these species live in tropical forests also called rain forests. The clearing and burning of these forests results in the disappearance of 4,000–6,000 species each year because of habitat destruction. Massive clear-cutting in other regions, such as the northwestern United States, is also destroying many indigenous plant and animal species. About 60,000 acres of ancient forest are timbered per year for export. Studies have shown that many species crowd into relatively small areas of a forest habitat so that clear-cutting often threatens survival of a species.

The "greenhouse effect" may pose an even greater effect on biodiversity than habitat destruction. Global warming is predicted to threaten many plant and animal species, particularly in temperate and polar regions where they will be unable to migrate fast enough to insure their survival.

The consequences of a rapid reduction in biodiversity are far-reaching. Natural selection has enabled species to adapt to even severe changes in habitat and climate. Natural selection, however, relies on an extensive gene pool so that a reduction in existing animal and plant species limits the possibility of adaptation to environmental stresses. Destruction of plant life also decreases our access to possible medicines, food sources, fiber, and petroleum substitutes. For example, about 75,000 plant species of the rain forests have edible parts, and potent medicines for Hodgkins disease and acute lymphocytic leukemia have been derived from tropical plants.

IDEAS TO EXPLORE

- What are the causes of the unprecedented decline in biodiversity?
- What estimates on the rate of decline have been proposed? How have they been formulated?
- Identify regions and ecosystems in particular where plant and animal species are at risk. What are some solutions?
- What are the consequences of mass extinction?

SOURCES

Scientific American, Sept. 1989, p. 108.
E. O. Wilson

Science, vol 241, Sept. 16, 1988, p. 1441.
R. M. May

SYMBOL KEY	
●	Topic of average difficulty
Δ	Long-term assignment
#	Project requires special facilities, equipment or supplies
○	Large public or college library required
*	Safety precautions required
+	Highly Technical; specialized knowledge required

DEPLETION OF FISH IN U.S. COASTAL WATERS ☉

The survival of many fish species is presently being threatened by pollutants, overfishing, and degradation of habitats and wetlands used for spawning and migrating. According to a recent report by the National Fish and Wildlife Foundation only 15 percent of those fish species caught 200 miles within the U.S. coastline are at acceptable levels while 14 percent, including the Atlantic salmon, Pacific Ocean perch, and California halibut, mackerel, cod, haddock and flounder, may be depleted in the near future. Pesticide contamination of estuaries which support marine life is thought to be responsible for the alarming decline of such species as striped bass in San Francisco Bay, and in Albemarle and Pamlico Sounds of North Carolina. The Atlantic salmon is gone from New England except for Maine because its upstream migration for spawning has been blocked by dams. Overfishing in New England has changed the population of fish stock so that those fish that are better able to survive are replacing more valuable species. Sole, yellowtail, flounder, fluke and other valued stocks which once represented 65% of all fish stocks of the Georges Bank now are reduced to 25%. Although in the 1960's Chesapeake Bay in Maryland saw a devastating decline in its oyster population, now only 1% of its former level, the rockfish and shad populations are returning after eight years of efforts by several states and the U.S. Environmental Protection Agency to reduce pollution in the bay.

IDEAS TO EXPLORE

- How have human activities such as overfishing, pollution emissions, and coastal development affected marine ecosystems?
- Have birds and marine mammals been affected by the sharp decline of fish in coastal waters?
- Why are estuaries so vital to the survival of many fish species?
- What steps were taken to help the rockfish recover in Chesapeake Bay?

PROJECT

Effects of Human Activity on a Coastal Region [•]

Select a coastal region or ecosystem and determine the effects of human activity on the survival of its inhabitants. Use graphs and charts to summarize data. Conservation organizations and government agencies may help provide you with information. What are possible solutions based on previous research and practical experience?

SOURCES

Center for Marine Conservation, Washington, DC

National Fish and Wildlife Foundation

National Marine Fisheries Service

U.S. and State Environmental Protection Agencies

WILD FLOWERS ⊡

Many species of indigenous wild flowers have been replaced by flowers that residents find more appealing. Indigenous wild flowers, however, generally play important roles in the health of ecosystems.

PROJECT

Local Wildflowers ⊡

Find out the various species of wild flowers or plants that have grown in your area. Determine their contribution to their ecosystem such as providing food for wildlife, adding nutrients to the soil, preventing soil erosion, deterring pests, etc. Determine the conditions under which your plant or plant species best thrive. What might be the effects of diminishing the number of wild flowers on their ecosystem? Of replacing them with foreign varieties of flowers?

SOURCES

Local horticultural societies and botanical gardens.

CROSS-POLLINATION OF PLANTS

PROJECT

Breeding a New Iris Plant 

You can breed a new iris plant by removing pollen from the stamen of one iris plant and brushing it on the stigmas of another. The iris plants, however, must be diploids (two sets of chromosomes in their cells) so that you must check with your supplier to be sure your parent iris plants are suitable.

SOURCE

Organic Gardening, May/June 1991, p. 88.

SYMBOL KEY	
●	Topic of average difficulty
Δ	Long-term assignment
#	Project requires special facilities, equipment or supplies
○	Large public or college library required
∗	Safety precautions required
+	Highly Technical; specialized knowledge required

DEFENSE MECHANISMS OF PLANTS ● ○

Recent studies have shown that plants have developed biochemical mechanisms to survive environmental stresses such as drought, pest infestation, and decreased light. For example, wilted plants release abscisic acid, a hormone that causes the synthesis of protective proteins that resist drought. Plants whose leaves are being attacked by pests have been shown to release chemicals that make them less appealing in order to repel the infestation. The cells on the surface of the leaves of some plants become more convex in the shade in order to capture more light for photosynthesis.

IDEAS TO EXPLORE

- What is known about how plants defend against environmental stresses such as drought, pests, and lack of adequate light?
- What biochemical changes have been shown to explain these defense mechanisms?

PROJECT

How Plants Respond to Decreased Light and Water ⊡

Select a houseplant that you can provide with recommended amounts of light and water. Use as many plants as possible for your study. First, provide the plants with appropriate care so that they adjust and clearly are thriving. Then divide the plants into three groups. Group one should continue to receive recommended amounts of light and water. Group two should be exposed to less light. Group three should be given less water. Observe the changes in the plants in response to these environmental stresses. Does the rate of growth change? Do the leaf sizes differ? Are there color differences?

SOURCES

Science News, August 11, 1990, p. 85, 86.
R. Donohue et al.
University of Wyoming, Laramie, Wyoming

R. Quatrano et al.
Oregon State University
Corvallis, Oregon

AN ECOLOGICAL STUDY

PROJECT

Understanding the Ecology of an Area ● ○ Δ

Identify the plant communities in an undeveloped area such as a nearby park or any other region you enjoy. Research the history of the area in order to understand the ecology. Was it once a mature forest cleared by logging and forest fires associated with lumbering? Was the soil irrevocably altered from the land disturbance so that different vegetation types thrived? Use photographs, sketches, samples, and maps to confirm the kind and location of plants and vegetation and also to support your study. Are there wetland communities? Coniferous forests? Northern hardwoods? Shrub thickets? What successional changes have occurred? Why?

SOURCES

A textbook on ecology will give you necessary background information. If you have selected a national or state park, information will be available at the Visitors Center. If you are interested in a region other than an official park or conservation area, try the Chamber of Commerce of the state as well as the ecology or biology departments of universities near the area.

BLUEBIRDS OF NORTH AMERICA ⊡

Bluebirds have long been favorites of both professional ornithologists and lay bird-watchers perhaps because of their attractive plumage and sweet song. Their population has diminished, however, due to loss of habitat as well as natural disasters. As a result, within the last decade individuals and groups have constructed bluebird nesting boxes and have taken an active role in increasing their numbers with significant success. Much of this work has been supported by the North American Bluebird Society which also sponsors research on the study of the ecology of North American bluebirds and publishes a quarterly journal called *Sialia.* Although bluebirds have been studied mainly on the North American continent they are known to occur in Mexico and northern Central America. The bluebird's generic name is *Sialia* and consists of three species; the Eastern, Western, and Mountain Bluebird.

IDEAS TO EXPLORE

- What is known about the behavior of the North American bluebird?
- Why has the bluebird population diminished so markedly? Have other song birds been affected?
- Is the decline in bird populations a measure of severe environmental destruction threatening our own survival?
- What measures have been taken to help the bluebird and other bird species survive?

PROJECT

Encouraging Bluebird Survival •

The North American Bluebird Society has chapters throughout the U.S. participating in various studies. Write the organization to locate a chapter in your vicinity to learn of ongoing projects which you might participate in or other ideas from members which might benefit the survival of bluebirds in your area.

SOURCES

North American Bluebird Society
Box 6295
Silver Spring, MD 20916-6295

SYMBOL KEY	
●	Topic of average difficulty
Δ	Long-term assignment
#	Project requires special facilities, equipment or supplies
○	Large public or college library required
*	Safety precautions required
+	Highly Technical; specialized knowledge required

THE DISAPPEARING SONGBIRD ⊡

In the 1960's Rachel Carson alerted us to the eventual arrival of a "Silent Spring" if environmental pollution and destruction of habitat were allowed to continue. Recent data on the declining populations of migrating songbirds strongly support her predictions. Some 250 species of migrating songbirds that breed in the temperate zone of North America and winter in Mexico, Central America, and the Caribbean are declining in numbers. Warblers, thrushes, and flycatchers are among those songbirds whose tropical forests are being destroyed. Bachanan's warbler may have become extinct due to loss of winter habitat.

IDEAS TO EXPLORE

- What environmental changes have occurred in the last 100 years to affect the migrating forest-dwelling songbirds?
- Have their breeding areas been markedly reduced or altered?
- Have their wintering ranges been changed by human activity?
- What are some possible solutions to conserve migrants in both the temperate and tropical zones?

PROJECT

Studying Local Songbirds •

Identify migrating songbirds that reside in your area or perhaps in a park nearby. Determine where they winter and where they breed. Examine what is known about their life cycle. Who are their predators? Where and how do they nest? How has human activity affected their life cycle? their success in reproduction in particular? Contact a local bird watcher's organization which may provide you with information and help you observe and record important aspects of the songbirds in your area. Also try a department of wildlife ecology or biology department of a nearby university. What conservation strategies are needed?

SOURCES

"Where Have All the Birds Gone?"
Essays on the Biology and Conservation of Birds that Migrate to the American Tropics, 1989
John Terbough
Princeton University Press
Princeton, NJ

The Sciences, Jan/Feb 1991
R. J. O'Connor

THE SPECIAL PROPERTIES OF CELL-WATER ⊡

The properties of water within living cells differ markedly from those of water outside of living cells such as water in a lake or in a cup of tea. Cell-associated or vicinil water has a lower density, and a greater heat capacity and viscosity than ordinary water. The ratio of potassium ions to sodium ions in cell water is 20:1 while in plasma the ratio is reversed. The unique properties of cell water are thought to be responsible for the ability of cells to survive even after severe water loss.

IDEAS TO EXPLORE

- How does solvation explain some of the differences between vicinil and ordinary water? How do these different properties help cells to survive and function?
- How can the properties of vicinil water explain why the internal temperature of mammals is around 98.6°F?

SOURCES

The Sciences, Sept./Oct. 1989, p. 38.
W. Drost-Hansen and J. Lin Singleton
University of California, Davis

Cell-Associated Water, W. Drost-Hansen and J. S. Clegy, ed, Academic Press, 1979

HUMAN GENE THERAPY ⊡

Gene therapy is based on the assumption that treatment of genetic diseases is possible by replacing or supplementing the defective gene. During the last decade several methods have been developed to introduce foreign normal genes into mammalian cells. It has been shown that in some cases genetic function can be restored by the addition of genes into nonspecific sites of the genome without removal or correction of the mutant gene. Gene transfer has been achieved through both physical and chemical methods. Virus vectors in particular have proved to be most efficient in delivering nucleic acids into mammalian cells *in vitro* followed by implantation of the genetically modified cells into a suitable organ. *In vivo* delivery of genetically modified cells is less developed but may be more effective. Diseases which one day may be treated with gene therapy include bone marrow diseases, central nervous system disorders, and cancer.

IDEAS TO EXPLORE

- How can genes be transferred? What are some problems involved in gene transfer?
- What are some examples of *in vivo* gene delivery?
- What are some ethical considerations concerning the development of gene therapy?

SOURCES

American Journal of Medicine, vol 83, 1987, p. 291.
M. Cline

Nature, vol 336, 1988, p. 348.
S. Mansour, et al.

Science, June, 1989, p. 1275.
T. Friedmann

SYMBOL KEY	
●	Topic of average difficulty
Δ	Long-term assignment
#	Project requires special facilities, equipment or supplies
o	Large public or college library required
*	Safety precautions required
+	Highly Technical; specialized knowledge required

GENETICALLY ENGINEERING PLANTS

The development of gene transfer methodologies for plants represents one of the most dramatic advances in agricultural technology. Genetically engineered plants are being developed to improve crops and to conserve the environment. Derivatives of the plant pathogen *Agrobacterium tumefaciens* have been used successfully to introduce genes into plants and plant cells as well as some physical means such as microinjection and using a particle gun. Some goals of genetic engineering for crop improvement include developing (1) herbicide-tolerant crops so that the more toxic herbicides and less biodegradable ones will not be needed for weed control; (2) insect-resistant plants; (3) disease-resistant crops. Transgenic tomato, tobacco, and potato plants have shown significant resistance to some problematic viral infections.

IDEAS TO EXPLORE

- What techniques are used in gene transfer of plants? What are some examples?
- What issues will affect the actual introduction of genetically engineered plants into the field on a large scale?
- What are some other applications of gene transfer in plants?

SOURCES

Science, vol 223, 1984, p. 496.
R. Horsch, et al.

Science, vol 244, 1989, p. 1293.
C. Basser and R. Fraley.

SYMBOL KEY	
●	Topic of average difficulty
Δ	Long-term assignment
#	Project requires special facilities, equipment or supplies
○	Large public or college library required
∗	Safety precautions required
+	Highly Technical; specialized knowledge required

4

CHEMISTRY

DETERMINATION OF VITAMIN C CONTENT OF FRUIT JUICES

Vitamin C, or ascorbic acid, is a necessary, water soluble nutrient that occurs in many fruits and vegetables. Recent studies have shown that its concentration decreases as foods sit, even if they are refrigerated, over a period of a few days.

PROJECT

Measuring the Amount of Ascorbic Acid in Juice

The amount of ascorbic acid in juice can be measured by a simple analytical technique which relies on the ability of ascorbic acid to be oxidized even by a fairly weak oxidizing agent such as iodine. Ascorbic acid is the only common component of plant or animal tissue that is readily oxidized in an acid or neutral solution so that oxidation is a very specific process. The reaction is

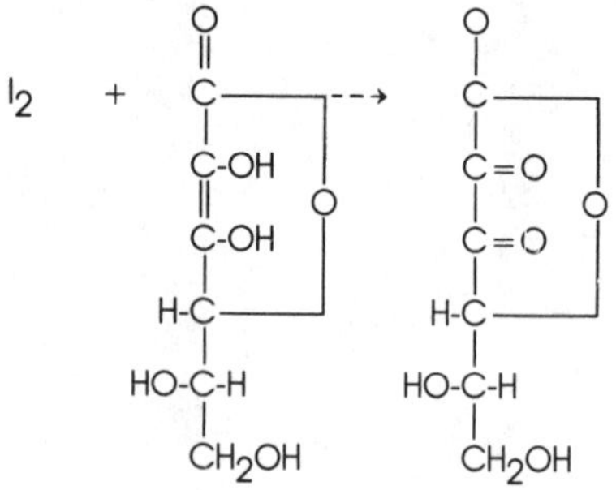

```
              O                  O
              ||                 |
I2    +       C--------┐ --->    C--------┐
              |        |         |        |
              C-OH     |         C=O      |
              ||       O         |        O
              C-OH     |         C=O      |
              |        |         |        |
            H-C--------┘       H-C--------┘
              |                  |
           HO-C-H             HO-C-H
              |                  |
              CH2OH              CH2OH
```

Ascorbic acid is titrated with I_2. The red color of I_2 disappears as it reacts. After all of the ascorbic acid is oxidized, the addition of more I_2 will result in a red color. The endpoint can be made more obvious by the addition of starch solution to the original sample. The starch will combine with the excess I_2 to form a complex with an intense blue color.

I_2 + starch ⟶ (starch-I_2) complex
red-brown — deep blue

Your project could compare the amounts of vitamin C in various juices and juice blends. You could also determine how much vitamin C remains in a container of juice that is opened and left in a refrigerator for 1 day, 2 days, etc. or how much remains if a juice container has been exposed to air and left at room temperature. Is the rate of oxidation of vitamin C temperature dependent? Recent studies on fresh fruits and vegetables have shown that vitamin C disappears rapidly so it is recommended that these foods be purchased fresh every couple of days.

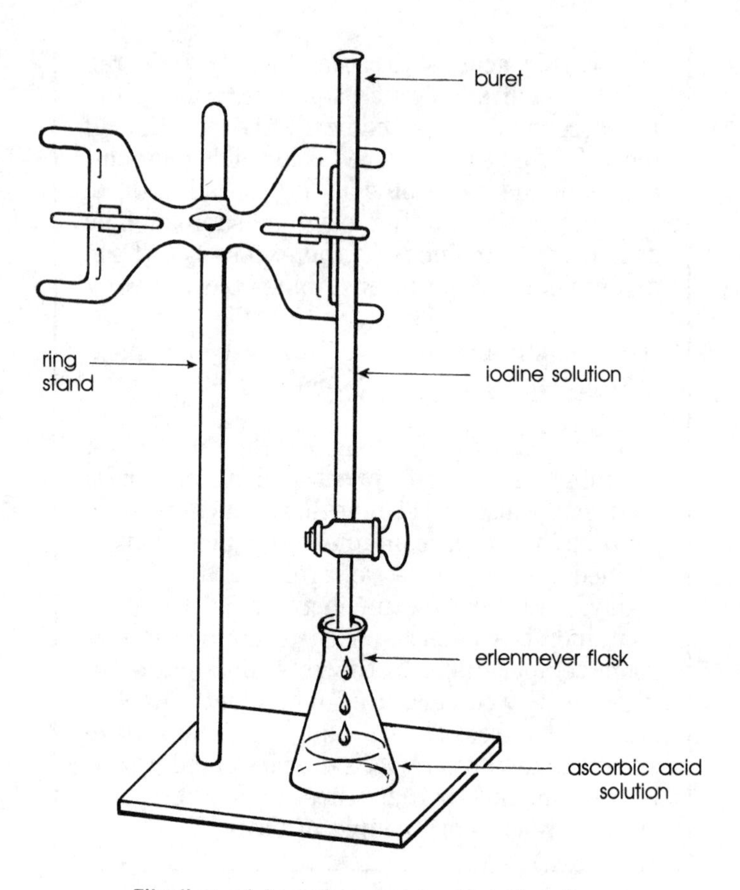

Titration of Ascorbic Acid with I_2 Solution

SOURCES

Laboratory Experiments in Chemistry.
West Publishing, 1990, p. 227.
A. D. Baker et al.
St. Paul, MN

ACTIVATED CARBON IN WATER PURIFICATION ● ○

Activated carbon is used to remove primarily organic contaminants from water and air. It is presently considered an effective treatment process for drinking water and is used by municipalities as well as by home owners to improve water quality. Activated carbon is produced from a variety of sources including coal and pits using different processes that determine its adsorption characteristics.

IDEAS TO EXPLORE

- How is activated carbon made? Why do its adsorption characteristics depend on the process used to prepare it?
- What contaminants are effectively removed by carbon filters?
- How is adsorptive capacity measured?

PROJECT

Effect of Saturation on Carbon's Ability to Adsorb Organic Contaminants ●

Many homeowners use activated carbon cartridges to clean water before it enters the house or they attach filters to individual faucets. These latter carbon filters can be purchased in most hardware stores and are fairly easily installed. They are thought to be effective in removing trihalomethanes (THMS) like chloroform, a byproduct of chlorination of municipal water, as well as other organic compounds. Chloroform is a known carcinogen and has been shown to be a universal problem of chlorination. As carbon, however, becomes saturated its ability to remove organic contaminants decreases. Your project can measure this decrease in the adsorption of organic substances with saturation.

You can use commercial carbon filters or make your own. Attach carbon filters to six or more faucets that have similar useage patterns and receive water from the same source. These faucets can all be in your house or some can be in the house of your neighbors. Remove one or more filters after the passage of each month up to about 6 months, the recommended lifetime of many commercial filters. You now have at least six carbon filters whose saturation levels differ. You can now measure their ability to adsorb organic compounds by exposing them to a test solution containing a food dye of your choice.

Dissolve enough food dye in one liter of water to make a test solution whose color is fairly intense. Also prepare a series of standard solutions whose concentrations are lower than that of the test solution. Filter about 25 mL of the one liter test solution through each of the partially saturated filters and collect the effluents. Estimate the concentrations of the effluent concentrations by comparing with the standard solutions. If the intensity of color of the effluent solution is between two of the standard solutions then so is the concentration of the effluent. Use test tubes to contain the solutions for the comparison. Once you have estimated the concentration of the effluent solutions you can then calculate an estimated percentage of dye adsorbed by each filter:

$$\% \text{ adsorbed} = \frac{\text{concentration of effluent}}{\text{concentration of test solution}} \times 100\%$$

SOURCES

"Occurrence and Removal of Volatile Organic Chemicals from Drinking Water," 1983
V.L. Snoeyink, American Water Works Association Research Foundation, Denver, Colorado

"Granular Activated Carbon," 1989
R. Clark and B. W. Lykins, Jr.
Lewis Publishers, Inc.
121 South Main Street, Chelsea, Michigan 48118

American Water Works Association
6666 West Queing Avenue, Denver, Colorado 80235

PROJECT

Measuring Heats of Reaction with Calorimeters ● ○

A calorimeter for measuring heats of reaction at atmospheric pressure (enthalpy changes, ΔH) can be constructed from a Styrofoam coffee cup for reactions in aqueous solution or from a thermos bottle with a wide mouth for hot or nonaqueous solution reactions. The cover of your "coffee cup" calorimeter or thermos bottle should be loose-fitting so that your reaction is open to the atmosphere. Insert a thermometer and stirrer through the cover and into your reaction mixture. By measuring the change in temperature of the solution caused by the reaction, the heat of reaction can be calculated. Styrofoam absorbs very little heat so that the solution itself will absorb virtually all of the heat evolved or contribute all of the heat absorbed.

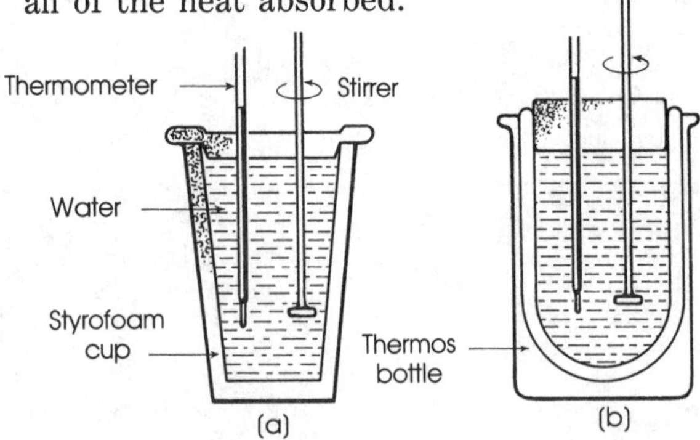

(a) The "Coffee-Cup" Calorimeter
(b) A Thermos Bottle Used as a Calorimeter.

SOURCES

Any college general chemistry textbook such as S. R. Radel and M. H. Navidi, *Chemistry*. West Publishing Company, 1990, p. 231.

PROJECT

A Demonstration of the Semipermeability of Biological Membranes ● ○

Semipermeable membranes are a common aspect of many biological processes. Ovolecithin liposomes can show semipermeability by using a chemical that can permeate the liposome where it will react with hydrogen peroxide to form a colored product which will remain within the membrane. The chemical entering the liposome could be ABTS (2,2-azino-di[3-ethyl benzthiazoline]-6-sulfonic acid), a green organic compound which in the presence of horseradish peroxidase (HRP) will undergo oxidation by hydrogen peroxide, producing the blue $ABTS^{+}$ cation. The cation, unlike the reactant and catalyst (HRP), cannot cross the membrane. Consequently, a blue pellet forms along with a colorless supernatant solution upon centrifugation. Addition of a detergent before centrifuging the colored suspension will disrupt the membrane so that a blue pellet cannot be isolated. The experiment is presented and described by A. Fimer in *Chemical Education*, vol 62, p. 89. If you select this project, in addition to learning about semipermeability of membranes, you also will learn about an important experimental technique, since liposomes are often prepared and used in research. You can enlarge upon this idea by using other chemical systems which you may discover from a literature search on semipermeable membranes. The needed chemicals and supplies can be obtained from chemical or scientific supply companies, some of which are provided under Resources.

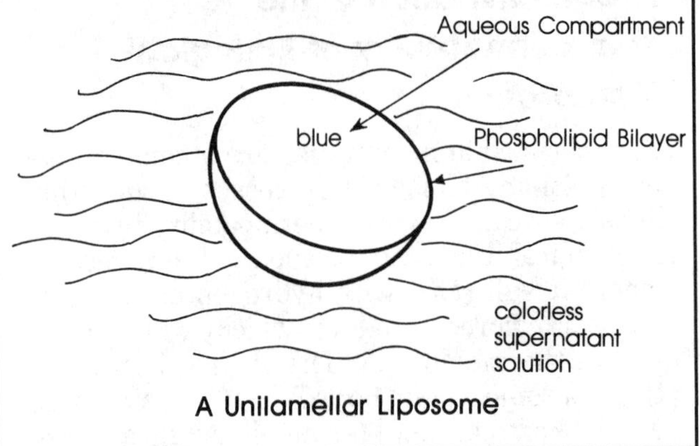

A Unilamellar Liposome

SYMBOL KEY	
●	Topic of average difficulty
Δ	Long-term assignment
#	Project requires special facilities, equipment or supplies
o	Large public or college library required
*	Safety precautions required
+	Highly Technical; specialized knowledge required

FATE OF CHEMICALS IN THE ENVIRONMENT + Δ

Today, approximately 60,000 chemicals are in commercial use, some of which are thought to pose potential risk to the environment and to human health. When these chemicals are released into the environment through atmospheric emissions, released into a water system or through a landfill their fate must be determined. For example, when a substance is released into the atmosphere some will enter a natural water system through precipitation, concentrations depending on volatility and aqueous solubility. Likewise, many organic chemicals released into a water system will also enter the atmosphere, depending on the chemical's Henry's Law Constant. Not only do substances distribute themselves between phases, but they may also undergo change in structure due to photolysis, bacterial action, or chemical reactions. These endproducts must be identified in order to determine the impact of the original substance on the environment. DDT for example, becomes DDE.

IDEAS TO EXPLORE

- What are the chemical and physical processes that chemicals undergo once released? Give examples of specific pollutants and their fate in the environment.
- For a more limited topic select a class of compounds, e.g., herbicides, organic solvents, or heavy metals and trace what is known about their interactions in the environment.

PROJECT

Predicting the Fate of a Specific Chemical +

Computer modelling programs are available for PC's. Because they tend to be expensive, check with the Environmental Science Departments of universities to see if a copy may be available. You could use a program to predict the fate of a particular chemical being released into the environment from a specific source. Be sure to report the uncertainties associated with the program you use and the underlying reasons for the uncertainties.

SOURCES

"Fate of Chemicals in the Environment," R. Swann, and A. Eschenroeder, ed. American Chemical Society Symposium Series 225

Consult Environmental Chemistry textbooks for basic information and journal references.

SYMBOL KEY	
●	Topic of average difficulty
Δ	Long-term assignment
#	Project requires special facilities, equipment or supplies
o	Large public or college library required
*	Safety precautions required
+	Highly Technical; specialized knowledge required

DIOXINS: THE CONTROVERSY CONTINUES

Dioxins have been described by the U.S. EPA (Environmental Protection Agency) as "one of the potentially most dangerous classes of contaminants in the environment." All dioxins have the structure

Dioxins

in which chlorine atoms occupy some of the numbered positions. The most toxic dioxin is 2, 3, 7, 8- tetrachlorodibenzo-p-dioxin (TCDD) which is considered "the most potent animal carcinogen ever tested." Related compounds are chlorinated furans which are thought to be toxic as well.

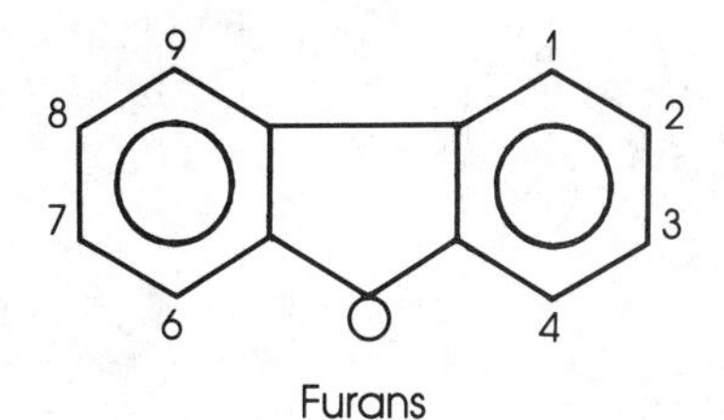

Furans

Furans often appear with dioxins. Dioxins and furans are released into the air by incineration of municipal and hazardous wastes. They are also formed in the commercial production of chlorinated chemicals such as the herbicide Agent Orange. The exposure of veterans of the Vietnam War to Agent Orange has been linked to an increased risk of cancer and reproductive and immune-system defects. Recently, it has been shown that the general population may be exposed to a much greater concentration of dioxin than previously thought. Dioxins have been found in chlorine-bleached paper products such as coffee filters, disposable diapers, paper towels, facial tissues, cereal boxes, newspapers, and even milk cartons.

IDEAS TO EXPLORE

- Dioxin is unquestionably a potent animal carcinogen and teratogen. Why is there so much controversy surrounding its risk to humans?
- Researchers at the University of Toronto found some 250 chlorinated compounds, some of which are dioxins and related furans produced from chlorine bleaching of pulp. Can these compounds migrate out of paper and into the food they contain? What are possible health risks?

PROJECT

Exposure to TCDD and Other Dioxins ● ○

Although there are several known sources of exposure to dioxins, researchers generally consider contaminated foods as the most significant source for the general population. Many studies have been reported which identify contaminated food sources. You can use this reported data to assess the exposure of volunteers from your community to dioxins. Record the consumption patterns of family members, friends, and neighbors for a month, for example, and use reported data to estimate their dioxin intake. The average daily intake in the U.S. is 1 pg/kg of body weight. Depending on the size and nature of your cohorts (volunteers), your data may determine if family members have similar exposure to dioxin or if members of the same age group have similar exposure. Is one age group at more risk than another?

A few known sources of dioxin are given below. To find other sources and estimates of concentrations you must consult the literature and you may need to gather information from EPA reports.

- Fish in contaminated lakes (such as the Great Lakes) because of bioaccumulation could contribute significant quantities to total exposure.

- Root crops such as potatoes, carrots, and onions develop TCDD levels exceeding levels in the soil.
- Foods and beverages packaged in paper products.
- Dioxins in coffee filters and milk may increase exposure of heavy coffee drinkers and children considably.
- Foods prepared in a microwave may also contain dioxins which migrate out of the paper tray.

SOURCES

Science News, vol 135. p. 104

Chemosphere, July 1988
Donald Mackay et al.

5

EARTH SCIENCES

VOLCANOES AND EARTHQUAKES [• Δ]

The latter part of this century has seen the eruption of volcanoes that have been dormant for hundreds of years as well as a significant increase in the number of earthquakes. In May 1991, Mt. Pinatuba, inactive for 600 years, erupted in the Phillipines and Mt. Unzen, dormant since the 1700s, erupted in Japan. Both of these volcanoes are along the "Ring of Fire," where the Pacific plate is sinking along with other parts of the earth's surface.

IDEAS TO EXPLORE

- What causes volcanoes to erupt? What causes earthquakes? Are they related? Can they be predicted?
- What scientific techniques are used today to monitor and measure these natural events?
- What are their short- and long-term effects on the environment?
- For a more limited report you could select to discuss in detail what is known about a particular mountain with a history of volcanic activity or a region that has been or is expected to be the site of earthquakes.

Volcanoes and Earthquakes in the United States

- Describe the major earthquake regions of the United States, giving the most recent data provided by seismologists studying earthquake activity in the area. Mt. St. Helens in Washington spewed particles and smoke into the atmosphere affecting weather patterns in other regions. What is known about this volcano? Are there others?

- Research volcanic or earthquake activity past, present, and future in your region or in another specific region in the United States.

PROJECT

Mapping Earthquake and Volcano Sites [•]

Construct a map of the United States showing sites of earthquakes or volcanoes in the past as well as possible activity in the future. You could use a system of symbols to indicate past activity as well as to indicate the probability for eruptions in the future (within 50 years, for example). Also show the presence of structures like nuclear power plants or natural formations like rivers or lakes that may markedly increase the destruction from an earthquake or volcano.

SOURCES

Begin by consulting a geology textbook such as *Modern Physical Geology*, 1991
G. Thompson and J. Turk
Saunders College Publishing

PROJECT

The Geology of Your State or Community [•]

Determine what geological forces formed the landscape of your state or community. Were mountains formed by collision of teutonic plates or erosion from a glacier? Was your state under an inland sea? Collect rocks to support the geological history you present. Are they sedimentary, igneous, or metamorphic? In some areas of the country fossils can be readily found or obtained. How do they support the geological history you present?

SOURCES

Museums and Nature Centers in various states can provide information. Also, you can try geology departments of local universities if you are unable to obtain sufficient material from a library.

SYMBOL KEY	
•	Topic of average difficulty
Δ	Long-term assignment
#	Project requires special facilities, equipment or supplies
○	Large public or college library required
*	Safety precautions required
+	Highly Technical; specialized knowledge required

6

ENVIRONMENTAL SCIENCE

THE EFFECTS OF "GREENHOUSE GASES" ON GLOBAL WARMING • Δ

The combustion of fossil fuels such as petroleum and natural gas, largely by industrialized nations, as well as clearing of tropical rain forests has led to a substantial increase of carbon dioxide in the atmosphere. The concentration of CO_2 has increased from 280 parts per million (ppm) in 1870 to more than 345 ppm at present. Carbon dioxide is known as a "greenhouse" gas because it traps radiation released by the earth, thus preventing heat from escaping the earth's atmosphere. (CO_2 absorbs in the infrared spectrum and reradiates heat downward from the atmosphere, warming the earth as a result.) Various estimates have been made regarding the rate of warming of the atmosphere and its effect on ecosystems.

Chlorofluorocarbons, methane, and nitrous oxide gases are also greenhouse gases whose emissions must be reduced.

IDEAS TO EXPLORE

- What are the major greenhouse gases and their sources?
- What are some of the predictions regarding an increase in temperature? other climatic changes? How are they determined?
- What may be the effect of the warming of the earth's climate on agriculture? on health?
- What are some possible solutions?

Is Global Warming Really Happening?

- The increase of CO_2 in the atomosphere due to human activities is being blamed for present and future climatic changes. Predictions vary portending dire consequences to the human community to very little effect relative to natural climatic changes.

IDEAS TO EXPLORE

- Why are scientists debating over whether global warming is occurring from the documented rise in CO_2?
- What is the indicator or indicators that global warming is occurring?
- What regions of the world, of the United States will be adversely affected by climatic changes due to greenhouse gases?

PROJECTS

1. Effects of Global Warming on a Specific Region •

Make your own predictions about the effects of global warming on your community, city, state, country or another area you select. Leaders of the Caribbean Islands are particularly concerned because of a predicted rise in sea level associated with global warming. Such a rise would decrease the size of the islands.

2. Effects of Global Warming on an Ecosystem [•]

Select an ecosystem and predict the effects of the warming of the atmosphere on the ecosystem.

3. Effects of Global Warming on a Specific Species [•]

Select a specific species of animal or plant and discuss effects of a rise in temperature.

SOURCES

Environmental Science and Technology,
April, 1990.
R. Lindzen

Science News, *36*, 359 (1990)

Environmental science and environmental chemistry textbooks will give you basic information as well as other references. Examples are *Environmental Science*, 1987.
J. Turk and A. Turk
Saunders College Publishing, U.S.A.

Environmental Chemistry, 1983
J. Moore and E. Moore,
Academic Press, NY

DEPLETION OF THE OZONE LAYER ● ○

A layer of ozone exists in the stratosphere between about 15 and 50 kilometers that absorbs much of the biologically damaging ultraviolet radiation emanating from the sun. Chemicals such as chlorofluorohydrocarbons (CFC's) and nitrogen oxides released into the environment have been shown to react with ozone. As a result, it is thought that production of these chemicals will lead to a reduction of the ozone layer which protects life on earth from the harmful UV rays. A seasonal hole in the ozone layer has been discovered above Antarctica causing elevated levels of UV radiation to reach polar life forms during these periods. The photosynthetic activity of phytoplankton at the bottom of the polar food chain has been shown to be markedly reduced by increased levels of UV radiation.

IDEAS TO EXPLORE

- What evidence is there that the ozone layer is in fact being depleted? What chemicals are thought to reduce the ozone level in the stratosphere? What reactions are thought to occur? What are these chemicals used for?
- Why is it important to maintain the ozone level in the stratosphere?
- Is skin cancer increasing in the United States?

- What are some other possible deleterious effects resulting from a depletion of the ozone level? What are some solutions?

PROJECTS

1. Effects of UV Radiation on Growth and Development of Plants • # *

Determine the effects of UV radiation on the growth and development of plants by using UV lamps. Measure the effect, if any, of UV radiation on the germination period for seeds planted on the surface of soil. Subject one group of plants to an elevated level of UV radiation, and monitor the growth of these plants relative to a control group receiving only ambient light. If several UV lamps are available, subject different groups of plants to different levels of UV radiation by using different UV exposure periods.

SYMBOL KEY	
•	Topic of average difficulty
Δ	Long-term assignment
#	Project requires special facilities, equipment or supplies
○	Large public or college library required
*	Safety precautions required
+	Highly Technical; specialized knowledge required

2. Effects of UV Radiation on Microorganisms • # *

The effect of increased levels of UV radiation on microorganisms in pond water can also be measured by using UV lamps. Examine measured samples of pond water under a microscope. Then subject these samples to enhanced levels of UV radiation. Observe the distribution of the microorganisms in each sample before and after radiation. Remember to use control groups.

Note: UV radiation is harmful; special glasses and a laboratory environment are suggested.

SOURCES

Environmental Science and Technology, vol 23, 1989, p. 1329.
C. Jackman

Science News, 10/14/89 p. 248, 10/28/89 p. 284

Nature, vol 315, 1985, p. 207
J. Farman et al.

NASA and the National Oceanic and Atmospheric Administration collect data and study ozone layers on an ongoing basis.

THE MYSTERY OF CLOUDS AND GLOBAL WARMING ⊡

Clouds are formed when surface water (mostly from the oceans) evaporates and condenses on dust, sea salt, bits of organic matter and other microscopic airborne particles. Eventually the water returns to the earth as rain or snow. At the center of the question of how clouds affect climate is the fact that clouds both heat and cool the earth. The cooling properties of clouds arise from the water as ice particles that reflect between 30 and 60 percent of the incident sunlight. Clouds also, however, exert a greenhouse effect by absorbing the heat eminating from the earth's surface and reradiating some of it back down. To complicate matters further, clouds appear in different forms determined by the weather conditions which create them. Stratocumulus clouds come in low, dense sheets which hover just above the ocean and have the effect of cooling more than they heat. Cirrus clouds are relatively thin and appear at 20,000 feet or higher, reflecting little sunlight and absorbing the earth's thermal radiation. They therefore warm more than they cool. Because of their complex effects, clouds represent a major obstacle in developing models to predict the climatic changes that the rapid rise in greenhouse gases may induce. As a result of the importance of their impact on weather patterns and global warming, clouds are presently being intensively studied.

IDEAS TO EXPLORE

- How are clouds formed and how do they impact on the radiation balance of the earth?
- What is known about the influence of clouds on weather? Why are they important in constructing models to predict future climate changes?
- Why is an understanding of clouds crucial in forcasting global warming?

SOURCES

The Sciences, May/June 1991, p. 36
W. Rossow

SYMBOL KEY	
●	Topic of average difficulty
Δ	Long-term assignment
#	Project requires special facilities, equipment or supplies
○	Large public or college library required
∗	Safety precautions required
+	Highly Technical; specialized knowledge required

EFFECTS OF TROPICAL DEFORESTATION • Δ

Tropical rain forests are being destroyed at an alarming rate. Already 50% of the total rain forests are gone. It has been estimated that 54 acres of tropical rain forest disappear every minute—that means an area the size of Florida and Maine combined is lost every year. These areas are cleared largely for agriculture. The effects on both the local and global environment are far reaching.

Local Effects of Tropical Deforestation

- The demise of forests means a loss of food, shelter, and other resources for local inhabitants. Environmental effects such as soil erosion, reduced rainfall, reduced ability of soils to hold water, increased frequency and severity of floods, and increased temperatures pose severe problems, especially to members of developing nations who struggle to survive.

IDEAS TO EXPLORE

- How are these environmental changes caused by clearing of the tropical forests? Give some specific examples.
- Can tropical forests be restored?

Global Effects of Tropical Deforestation

- Although more than half of the world's species inhabit tropical rain forests, we have studied, for example, only about 1% of the tropical plant life. Loss of habitat leads to extinction of species so that many unknown species may never be studied. Other far reaching consequences of deforestation include alteration in the earth's water cycle, heat balance, and climate, and emission into the atmosphere of CO_2, CH_4, N_2O, and CO. All of these compounds are greenhouse gases which are thought to contribute to global warming.

IDEAS TO EXPLORE

- Why is it predicted that tropical deforestation will alter the earth's climate?
- Why does deforestation cause the emission of greenhouse gases?
- What species have already been affected by the massive destruction of the forests? What others are in danger?

PROJECT

Effect of Deforestation on a Single Species [•]

Select one nation or area within a country and study one or more animal or plant species. Predict the effect of a decline in tropical forests and increase in exposure to humans on the survival of that species. What is the effect of a decline of forests on the food supply of that species? How will reproduction be affected? What about success in avoiding predators?

SOURCES:

Science News, vol 138, p. 40.

Environmental Science and Technology, April 1990, p. 414
R. Houghton

"The Fall of the Forest," Richard Monasteresky
U.S. EPA, Washington, DC

United Nations Food and Agriculture Organization

World Resources Institute (WRI)

Thomas Lovejoy of Smithsonian Institute

SYMBOL KEY	
●	Topic of average difficulty
Δ	Long-term assignment
#	Project requires special facilities, equipment or supplies
○	Large public or college library required
*	Safety precautions required
+	Highly Technical; specialized knowledge required

AIR POLLUTION • Δ

Outdoor air pollution is known to be harmful to humans, to cause damage and death to trees and agricultural crops, to cause poor visibility and corrode building materials, and to acidify rain water, affecting the health of lakes and rivers. The burning of fossil fuels by power plants and automobiles is largely responsible for outdoor air pollution. Major pollutants include sulfur dioxide, ozone, carbon monoxide, nitrogen oxides, lead, and particulate matter.

IDEAS TO EXPLORE

- What are the sources of major air pollutants in an urban area? rural area?
- How are pollutants measured and monitored?
- What are their effects on human health and welfare? on vegetation and animals? on the atmosphere, soil, and water? on materials and structures?
- What is photochemical smog?
- What are the "state of the art" air quality control methodologies and techniques?
- For a more limited topic choose one or two pollutants such as ozone or carbon monoxide and determine their sources, interactions with other components of the atmosphere, effects, methods of measurement and control technologies or discuss pollutants emanating from a specific source such as power plants, automobiles, nuclear reactors or waste disposal incinerators. Some hospital incinerators, for example, have recently been cited as significant sources of air pollution.

PROJECT

Measuring Air Pollution in Your Neighborhood •

Select one or two air pollutants that you can measure in your neighborhood. Both chemical and instrumental techniques have been developed for determining the concentration of air pollutants. Measure concentrations at different times of the day and different seasons if the project is long-term. Offer an explanation for your results by consulting the literature for information on known sources of these pollutants and their behavior in the atmosphere.

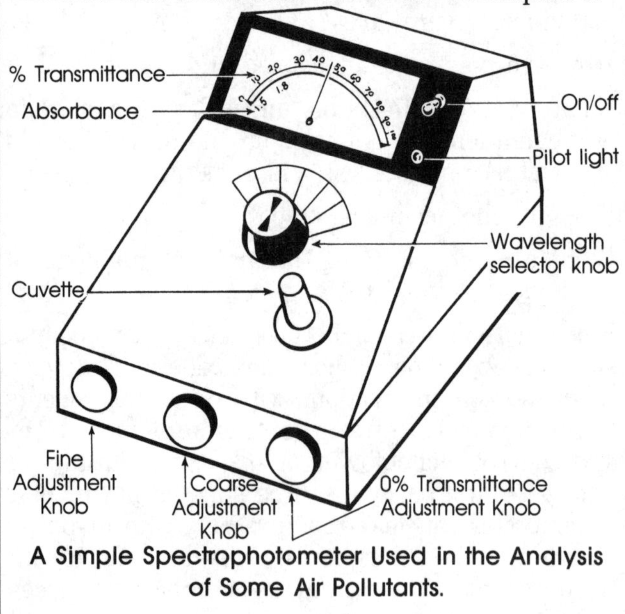

A Simple Spectrophotometer Used in the Analysis of Some Air Pollutants.

SOURCES

Environmental Science, 1987
Turk, J. and A. Turk.
Saunders College Publishing, U.S.A.

Scientific American, Sept 1990, p. 60
Crutzen P. and T. Graedel.

Environmental Science and Technology
April 1990, p. 442
Cortese, A. Tufts University

Write U.S. EPA, Office of Air Quality
Planning and Standards
Research Triangle Park, NC

Fundamentals of Air Pollution, 1984
A. Stern et al.
Academic Press, Inc., New York

SYMBOL KEY	
●	Topic of average difficulty
Δ	Long-term assignment
#	Project requires special facilities, equipment or supplies
○	Large public or college library required
*	Safety precautions required
+	Highly Technical; specialized knowledge required

PHOTOCHEMICAL SMOG • ○

The yellow-brown haze that appears above almost every modern city is called photochemical smog. The color is due primarily to NO_2, a brown colored gas which is released into the atmosphere from car exhausts. Smog results from many free radical reactions initiated by the cleaving of NO_2 into NO and O by sunlight. Nitrogen oxides are called primary pollutants while other undesirable pollutants like ozone (O_3) and peroxyacylnitrates (PAN) are secondary pollutants. These secondary pollutants are associated with the well known harmful effects of photochemical smog on the health of both people and vegetation.

IDEAS TO EXPLORE

- What causes smog?
- What are the harmful effects of smog?
- What can be done to reduce photochemical smog?

SYMBOL KEY	
•	Topic of average difficulty
Δ	Long-term assignment
#	Project requires special facilities, equipment or supplies
○	Large public or college library required
*	Safety precautions required
+	Highly Technical; specialized knowledge required

PROJECT

Photochemical Smog in a Specific Urban Environment

Select an urban area where smog is a problem. Determine the main sources of the primary pollutants in the area and trace the concentrations of secondary pollutants throughout the day. Do these concentrations depend on the season? Why? Suggest solutions to reduce photochemical smog based on your literature research. Concentrations of pollutants are generally measured by government environmental agencies in each city and are available to the public. Ozone levels in particular are often available on request.

SOURCES

Fundamentals of Air Pollution, 1984
A. Stern et al.
Academic Press, Inc., New York

Most environmental chemistry textbooks will provide you with basic information as well as references. An example is

Environmental Chemistry, 1983

J. Moore and E. Moore
Academic Press, Inc., New York

ALTERNATIVE ENERGY SOURCES ● ○ Δ

Presently, the United States depends primarily on non-renewable resources for its energy consumption. Fossil fuels, aside from being a limited resource, produce carbon dioxide and many other gases that are harmful to the environment and health. Nuclear energy uses fission reactions that produce highly radioactive wastes creating a major disposal problem. Chernobyl and Three Mile Island, among other incidents involving nuclear power plants, have emphasized serious shortcomings in the technology. Fusion, a much cleaner energy source, is presently being developed. Its introduction, however, will depend largely on the discovery of materials that can withstand the extremely high temperatures necessary for fusion reactions to occur. It is predicted that alternate energy technologies such as solar, geothermal, hydroelectric and wind energy will become more important in the 21st century.

IDEAS TO EXPLORE

- What renewable energy sources are being used today? What are advantages and disadvantages of these sources?
- Describe the state-of-the-art technology associated with each energy source. How can these technologies be integrated into society?
- What areas of research are being actively pursued to develop alternate energy sources?

PROJECT

Use of Alternate Energy Sources in Residential Home Design ● ○

Design a house in your community that would incorporate the "state-of-the-art" technology using an energy source or sources other than fossil fuels.

Solar Energy ● ○

The last decade has seen major innovations in harnessing the sun's energy. Many communities presently depend on solar energy for generating their electricity. More efficient photovoltaic cells have been developed as well as materials and processes that transform solar energy into useable forms.

IDEAS TO EXPLORE

- How can solar energy be used to reduce the consumption of fossil fuels?
- What technologies are available today? What areas of research are being pursued?
- How can these new technologies be adopted to decrease our dependence on fossil fuels and consequent release of air pollutants and carbon dioxide?

SOURCES

Scientific American, Sept. 1989, p. 136.
J. Gibbones et al.

American Society of Heating, Refrigerating and Air Conditioning Engineers (ASHRAE)

SYMBOL KEY	
●	Topic of average difficulty
Δ	Long-term assignment
#	Project requires special facilities, equipment or supplies
○	Large public or college library required
*	Safety precautions required
+	Highly Technical; specialized knowledge required

INDOOR AIR QUALITY • Δ

It was thought that exposure to polluted air could be controlled by government regulation of outdoor air. However, in the 1970's it became evident that by making buildings airtight to decrease energy usage another problem was created: unhealthy indoor air. Because 80% of Americans spend most of their time indoors, poor indoor air quality poses a major health concern. Known indoor air pollutants include radon, nitrogen oxides from gas stoves, sulfur dioxide from kerosene stoves, many pollutants from cigarette smoke, formaldehyde and other organic chemicals from insulation, glues and resins used in furniture and carpets. These pollutants have been associated with respiratory disease, cancer, and allergies.

IDEAS TO EXPLORE

- What are the major indoor air pollutants? What are their sources and suspected health effects?
- What is Sick Building Syndrome? What methods can be used to lower the concentrations of indoor air pollutants?

PROJECT

Effect of Plants on Indoor Air Pollution • # * +

Some plants have been shown to remove indoor pollutants. Determine the concentrations of sulfur dioxide and nitrogen oxides in different areas of your house. The concentration of these two air pollutants can be measured by spectrophotometric analysis. There are three basic parts to each analysis.

(1) Prepare a glass fiber filter impregnated with potassium carbonate. Expose this filter to indoor air for a four-week period.
(2) Prepare solutions for a calibration curve using Beer's Law.
(3) Prepare test solutions from the exposed filters, measure adsorbance using a Spectronic 20 spectrophotometer, and then determine the concentrations of your test solutions from your Beer's Law curve.

(For the details of spectrophotometric analysis consult: *Ideas, Investigations, and Thought: A General Chemistry Laboratory Manual*, 1980, Avery Publishing Group, Inc., Wayne, New Jersey. S. Kennedy et al., Dept of Chemistry, Hofstra University, L.I., N.Y.)

Measure radon concentrations with activated carbon cartridges that are readily available through mail order or local hardware stores. Call your state EPA to find out where these cartridges are available. Bring house plants into an area where pollution is measurable and where the plants will grow. Monitor the level of pollution every week. Select a small area with a low air exchange rate so that an effect, if there is one, can be observed.

SOURCES

Indoor Air Pollution
R. Wadden and P. Scheff.
1983, Wiley and Sons, Inc.

Scientific American, May 1988, vol. 258, p. 42.
Anthony V. Nero Jr.

Many articles have been published in *Science*, *Nature*, and the *New England Journal of Medicine*

U.S. EPA
Public Information Center
401 M Street SW
Washington, DC 20460
(202) 382-2080

The Inside Story: A Guide to Indoor Air Quality, Publication No. EPA/400/1-81-004.

EPA Report to Congress on Indoor Air Quality

SYMBOL KEY	
●	Topic of average difficulty
Δ	Long-term assignment
#	Project requires special facilities, equipment or supplies
○	Large public or college library required
*	Safety precautions required
+	Highly Technical; specialized knowledge required

INDOOR RADON AND LUNG CANCER ⊡

Uranium mine workers have higher rates of lung cancer than the general population. The decay products of radon-222 (^{222}Rn), a radioactive gas produced from uranium-bearing rock and soil, is thought to be the cause. Recently, radon has been detected in many American homes at concentrations existing in uranium mines and higher. It is currently advised that the concentration of radon gas in indoor air be less than 4 picocuries/liter. The Environmental Protection Agency has estimated that 10,000 or more deaths from lung cancer in the United States may be due to exposure to the decay products of radon in the home.

IDEAS TO EXPLORE

- What are the sources of radon?
- What factors contribute to the build-up of radon in the home?
- Have areas in your state been identified as containing homes with high radon concentrations?
- What evidence is there to suggest that exposure to radon and its decay products is a health risk?
- How can radon be prevented from entering your home?
- How is the concentration of radon measured?

PROJECT

Radon in Your Community • Δ

Collect data on the concentration of radon gas in homes in your area. The EPA strongly encourages people to test radon levels in their homes. Test kits are available in many hardware stores or can be easily obtained from companies registered with the EPA. See if there are areas where radon levels are high. What do these areas have in common? Are these houses lying above radium-bearing rocks or soil? Are they particularly well insulated so that radon gas accumulates?

SOURCES

The federal Environmental Protection Agency distributes free publications on radon. Also contact your state EPA and local newspapers for reports of radon gas concentrations in your area. The scientific journals *Science*, *Environmental Science and Technology*, and *Chemical and Engineering News* all contain articles on radon gas. Well known researchers in the field include William W. Nazaroff and Anthony V. Nero, Jr., both from Lawrence Berkeley Laboratory in California.

Radon and Its Decay Products in Indoor Air
edited by W. W. Nazaroff and A. V. Nero, Jr.
New York: John Wiley and Sons, 1988

STORAGE OF HIGH-LEVEL RADIOACTIVE WASTES ⊡

The United States generates high-level radioactive materials from nuclear power and weapon plants. The Department of Energy plans to locate a geologic repository for these wastes for at least 10,000 years at Yucca Mountain in northwestern Nevada.

IDEAS TO EXPLORE

- How much waste is actually involved? Where and how is it being stored now?
- What criteria must be met by the disposal site so that radionuclides are not released into the environment?
- What are some problems associated with using a geologic repository for nuclear wastes?
- What are some alternative methods of disposal? What are the advantages and disadvantages of each method or technology?

SYMBOL KEY	
•	Topic of average difficulty
Δ	Long-term assignment
#	Project requires special facilities, equipment or supplies
o	Large public or college library required
*	Safety precautions required
+	Highly Technical; specialized knowledge required

PROJECT

Nuclear Waste at a Particular Power Plant [•]

Select a nuclear power plant and find out the amount of waste generated and the composition of the wastes. Calculate how long it would take for the radionuclides to disintegrate. Where are the wastes presently being stored? What are possible solutions for the storage problem?

SOURCES

Site Characterization Plan; US Department of Energy; US Government Printing Office: Washington, DC, 1988, DOE/RW-0199

Geostatistical, Sensitivity, and Uncertainty Methods for Ground-Water Flow and Radionuclide Transport Modeling; Buxton, B. E. Ed; Batelle Press: Columbus, OH, 1989

ACID RAIN • Δ

Acid rain is caused by air pollution. It has made many lakes, streams, and ponds acidic causing major disruptions in the ecological balance of these water systems. Some lakes, for example, no longer support aquatic life. It is also suspected that some species of trees, especially those growing on mountain tops, have died or are dying from exposure to acid rain and acid clouds. Acid rain also contributes to the destruction of some building materials; the Greek parthenon has been damaged from exposure to acid rain and air pollutants in general. Agriculture may also be affected.

IDEAS TO EXPLORE

- What is thought to cause acid rain?
- In what areas of the nation is it most prevalent? Does it exist in other areas of the world?
- What are thought to be the effects of acid rain on ecosystems?
- What has been the response of the countries affected by acid rain? Have they been successful in mitigating the environmental damage?
- Select one of the environmental effects of acid rain given above and discuss the extent of the damage, possible mechanisms, and proposed solutions.

PROJECTS

1. Effect of Acid Rain on Germination of Bean Seeds • *

Unpolluted rain water is only slightly acidic due to dissolved carbon dioxide. The pH of rain water should be between 6 and 7. The pH of acid rain has been measured in some areas as low as 2 or 3. What is the effect of acid water on the germination and growth of plants? What is the effect of acid rain on the germination of bean seeds? Plant two or more groups of bean seeds using the same soil and exposing all seeds to the same conditions of temperature and light. Prepare solutions of water at different levels of acidity and expose each set of bean seeds to one of the solutions. Use dilute sulfuric or nitric acid to make up solutions.

2. Effect of Acid Rain on Plant Growth • Δ *

After germination monitor the rate of growth and the general health of the plants. You can also use solutions of the same pH prepared with nitric and sulfuric acids to determine whether the nature of the acid solution affects the germination and or growth of the plant. You can use plants other than beans or in addition to bean plants to see if acid rain has effects on some plants and not on others.

SOURCES

Consult your local nursery on germinating and growing bean plants indoors. Consult your science or chemistry teacher on making up the acid solutions and safely storing them.

3. Determine the pH of Local Water Sources •

Measure the pH of ponds, streams, and lakes in your area. Are these water systems too acid? Compare with pH values of water systems in other areas with similar geology. You may be able to borrow a portable pH meter from your school. If not, discuss with your science or chemistry teacher how to collect samples and then arrange to bring them into your school laboratory for measurement.

4. Seasonal Changes in pH

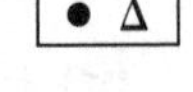

Measure the pH of local ponds, streams or lakes during the different seasons. What climatic changes could cause pH fluctuation?

5. Changes in pH of Rain Water • Δ

Collect rain water samples during different times of the day and year. Does the pH change? Why? Are the samples more acidic than unpolluted rain water?

SOURCES

Acid rain has been studied for at least ten years, and information is plentiful. Consult textbooks on environmental science or environmental chemistry for basic information. Articles appear in *Science*, *Scientific American*, *Environmental Science and Technology*, *Science News*, and *Chemical and Engineering News*.

Environmental Science, 1987
J. Turk and A. Turk
Saunders College Publishing, U.S.A.

SYMBOL KEY	
•	Topic of average difficulty
Δ	Long-term assignment
#	Project requires special facilities, equipment or supplies
o	Large public or college library required
*	Safety precautions required
+	Highly Technical; specialized knowledge required

NUCLEAR WINTER • Δ

Many scientists believe that a large scale nuclear war would have devastating long-term effects on the environment, causing drastic climatic changes and driving many plants and animals into extinction. It is thought that these severe environmental consequences would result from the enormous amounts of soot and dust spewed into the atmosphere from the fires and explosions caused by the nuclear warheads. The clouds of smoke would block the sun's rays, dropping temperatures worldwide and causing a nuclear winter. The ecological damage would threaten survival of homo sapiens.

IDEAS TO EXPLORE

- What data do scientists use to predict the effects of a nuclear war on the environment?
- Have natural events such as volcanoes affected climate?
- What are possible effects of massive amounts of smoke entering the atmosphere from the oil fires in Kuwait?

SOURCES

Science, vol 222, 1983, p. 1293.
P. R. Ehrlich et al.

Nuclear Winter, 1984
Mark A. Harwell
Springer-Verlag, NY

WATER POLLUTION ⊡

Many natural water systems such as lakes, rivers, oceans, and streams are polluted as a result of human activities. The harmful chemicals that find their way into these systems vary depending on the location of polluting sources. Water pollution can be caused in many ways such as from direct discharge of chemicals by industries, from urban and farm runoff, from the dissolving of gaseous chemicals in the air, from sewage disposal, etc.

Because the topic of water pollution involves so many different types of water systems and chemicals, it would be best for you to select one type of natural water system—lakes or rivers or even oceans—and discuss the various ways in which the water system is becoming polluted and the effects of the pollution on the ecosystem.

IDEAS TO EXPLORE

- What pollutants are present? How did they get there? How are they measured?
- What are the effects, both present and future, of these pollutants?
- How can the source of the pollution be stopped?
- What methods are available to remove the pollution?
- For a shorter paper you might either limit the region—lakes in the northeast United States, for example—or discuss one class of pollutants such as organic chemicals or heavy metals.

PROJECTS

1. History of Pollution in a Local Water System ○

Select a water system in your community or one close by that you have heard from news reports may be polluted. Trace the history of that pollution and its effect on the aquatic life and the ecosystem in general. What steps are being taken, or do you think should be taken, to reduce the pollution?

SYMBOL KEY

- ● Topic of average difficulty
- Δ Long-term assignment
- # Project requires special facilities, equipment or supplies
- ○ Large public or college library required
- ∗ Safety precautions required
- \+ Highly Technical; specialized knowledge required

2. Assessing the Health of a Local Water System •

Select a water system in your community such as a stream or pond and assess its health by collecting fish, insects, and microscopic plants. For example, some insects like mayflies and stoneflies tend to disappear from polluted streams while leeches and aquatic worms are numerous. Insects are abundant in healthy streams to feed birds and fish. Their absence, therefore, limits aquatic life higher in the food chain. You might also test the stream for chemicals such as nitrates and phosphates using test kits available from various chemical supply companies. The concentrations of nitrates and phosphates are measured by reaction with reagents which produce a colored product. The concentration of this product is then determined with a colorimeter. These test kits and complete instructions are sold by LaMotte Chemical Products Company (P.O. Box 329, Chestertown, MD 21620). The concentration of dissolved oxygen is a good indicator of the health of a stream. It tends to decrease with pollution. It can be measured with a portable monitor or with chemical tests.

SOURCES

Consult an environmental science book such as
Environmental Science, 1987
J. Turk and A. Turk
Saunders College Publishing, U.S.A.

HUMIC MATERIALS 

All natural water systems—oceans, lakes, rivers, streams—contain humic materials (large organic molecules resulting from the decay of plant life). In oceans these molecules serve as a source of carbon for plankton or bacteria. Sunlight splits the large molecules into smaller edible portions which microorganisms consume during the day and at night when no further production occurs. If there is or will be an increase in UV radiation from depletion of the ozone layer, how will concentrations of humic materials be affected?

SYMBOL KEY	
•	Topic of average difficulty
Δ	Long-term assignment
#	Project requires special facilities, equipment or supplies
o	Large public or college library required
*	Safety precautions required
+	Highly Technical; specialized knowledge required

PROJECT

Effect of UV Radiation on Humic Materials • # *

Determine the effect, if any, of increaed UV radiation on humic materials. Obtain humic acid from a chemical supply company such as Aldrich Chemical Company. Prepare ten solutions of humic acid having concentrations similar to those found in natural water systems. Have these solutions range from concentrations of 0.5 mg/L of organic carbon (typical of seawater) to 30 mg/L of organic carbon (similar to colored water from a swamp). Use a UV lamp to subject your solutions to increased UV radiation. (Be sure to keep control samples which are not exposed to the lamp.) Monitor any changes in concentrations of humic acid, using a spectrophotometer (visible range) or a colorimeter with exposure to UV radiation.

SOURCES

Nature, Oct. 19, 1989
K. Mopper
University of Miami

OIL SPILLS ⊡

In March 1989 the Exxon Valdez tanker spilled almost 11 million gallons of crude oil into Alaska's Prince William Sound. According to the Alaska Oil Spill Commission, oil discharges of this magnitude occur somewhere in the world once a year, on average, while spills of a million gallons occur once a month. The consequences of these spills are far-reaching; for example, an estimated 100,000–300,000 sea birds, thousands of marine animals, and hundreds of bald eagles perished as a result of the Exxon Valdez accident. Long-term effects from exposure of fish and wildlife to hydrocarbons are being studied but scientists are not optimistic based on laboratory results showing that these compounds are carcinogenic and may interfere with reproduction. The spill prevented subsistence fishing, hunting, and gathering by villagers living along the thousand miles of affected coastlines as well as disrupting commercial fisheries and tourism.

Much data has been acquired on oil spills from the Exxon Valdez event, involving the fate and transport of the thousands of different hydrocarbons in crude oil, environmental effects, environmental assessment, and methods and techniques used to contain and remove the oil.

IDEAS TO EXPLORE

- What was Alaska's response to the Exxon Valdez oil spill? Were the containment and recovery attempts effective? What new technologies were tried?
- How was the oil spill monitored? What models were used?

- What was the fate of the different kinds of hydrocarbons in the crude oil?
- Describe some of the immediate effects on the environment. What are predicted long-term effects? What is being done now to mitigate some of these long-term effects?

PROJECT

Impact of the Exxon Valdez Oil Spill [•]

Trace the fate and transport of the Exxon Valdez oil spill including the environmental impact to the sound and surrounding coastal areas. Maps and charts could be used to show the path of the spill and effects.

SOURCES

Environmental Science and Technology, vol 25, No. 1, 1991, p. 16 (five part series).
D. Kelso and M. Kendziorek

Try the Alaska Department of Environmental Conservation, Juneau, AK 99811

SOLID WASTE DISPOSAL: THE ERA OF RECYCLING ⊡

Historically, solid waste has been disposed of by depositing it in landfills. Landfills across the nation, however, are reaching their capacity, creating a crisis in solid waste disposal. These sites decrease property values in communities as well as present health risks. Toxic materials have leached out polluting underground and surface water. Methane, produced by bacteria, can migrate underground to nearby buildings and cause explosions. As a result many states have introduced regulations which increase markedly the standards of landfill design to reduce exposure of people to pollutants.

Incineration, conservation of resources, and recycling programs are other methods of solid waste management. More than 75% of the municipal solid waste in the U.S. can be recycled. Recyclable materials include paper, glass, metal, plastics, and vegetation by composting.

IDEAS TO EXPLORE

- What new landfill technologies have been developed to reduce the impact of solid wastes on the environment?
- What are the advantages and disadvantages of the most modern incinerator technologies?
- Describe the processes by which materials are recycled.
- How can these methods of disposal be incorporated into an integrated program of waste management?

PROJECT

Reducing Solid Waste •

It has been estimated that in five years the average American disposes of an amount of solid waste that weighs the same as the Statue of Liberty. Therefore, the reduction of disposal on an individual basis is necessary to effectively manage solid wastes.

Ask friends, neighbors, and family to record what they discard in a one- or two-week period. Which of these items could be recycled? Could the amount of material being discarded be reduced by resource conservation such as using canvas shopping bags, plastic reusable coffee cups, etc.? Do the generations differ in what and how much they discard?

SOURCES

Scientific American, vol 259, 1988, p. 36.
P. R. O'Leary et al.

SYMBOL KEY	
•	Topic of average difficulty
Δ	Long-term assignment
#	Project requires special facilities, equipment or supplies
○	Large public or college library required
*	Safety precautions required
+	Highly Technical; specialized knowledge required

THE PERSIAN GULF OIL SPILL ☐ • ○

Perhaps the most long lasting effects of the Persian Gulf War on humans, animals, and plants will be related to the devastation of the environment as a result of the oil spill in the Arabian Gulf and the Kuwaiti and Iraqi oil well fires.

The Gulf region has a high biological diversity with over 200 faunal species in the intertidal zone and 1,000 species occupying the coral reefs. Seagrass beds provide energy sources and shelter for species such as shrimp and fin fishes. The coastal wetlands are vital for more than 100 migratory bird species. About 10 million people in the Gulf region rely on the industries and activities associated with the Gulf.

IDEAS TO EXPLORE

- What ecosystems in particular in the Gulf are at risk as a result of the oil spill?
- What are the short-term effects? long-term effects?
- What species which reside in and visit the Gulf are at risk?

SOURCES

Fish and Wildlife News, Jan-Feb 1991, p. 1.
U.S. Department of the Interior

ENVIRONMENTAL EFFECTS OF THE PERSIAN GULF WAR ⊡

In March 1991 the Kuwaiti oil well fires were releasing 50,000 tons of sulfur dioxide, 100,000 tons of soot (carbon) and more than 800,000 tons of carbon as CO_2 daily. Some scientists believe that this will drastically affect agriculture as well as natural ecosystems in southern Asia. The Northern Hemisphere may be affected as well. It is clear that the air pollution in Kuwait is severe and perhaps life-threatening; temperatures have already decreased by 15 degrees Celsius.

IDEAS TO EXPLORE

- What are the predictions of various scientists about the global and local effects of the Kuwaiti fires?
- What does Carl Sagan mean by "self-lofting" when he refers to burning oil?
- Is a nuclear winter effect possible?
- What information is available now on the distribution of the pollutants and their effects? Does this data support the predictions offered by scientists?

SOURCES

Scientific American, May 1991, p. 17.

EFFECTS OF THE KUWAIT OIL WELL FIRES

Almost 600 Kuwaiti oil wells were set afire by Saddam Hussein's army during the Persian Gulf War. The known effects of these fires so far appear to be concentrated largely in the Persian Gulf region. They include an unseasonal cooling; Bahrain reported the coldest May in 35 years as a result of the soot clouds which block the incident sunlight. Currently the pollution is remaining close to the surface of the earth, posing a threat to all life in the Persian Gulf. High levels of particulates linked with high mortality rates in United States cities have been measured in Kuwait. Hospitals are reporting significant increases in new cases of respiratory problems, and it is predicted that 10 percent of people in southeast Kuwait will become sick. In this region crops and vegetation, as predicted, are being destroyed.

The data gathered at this point indicates that most of the soot and pollution is remaining in the lower atmosphere so that a nuclear-winter scenario in which global weather patterns would be markedly changed is not an immediate threat. Black snow and acid rain, however, have been reported over parts of Iran. Although EPA scientists feel that the global climate will not be affected by the oil well fires and that the Kuwaiti people are not at undue risk, other atmospheric scientists emphasize that it is too early to predict the total effects of the pollution on both a local and global level.

IDEAS TO EXPLORE

- How could the Kuwaiti oil wells affect the global climate?
- How could India's monsoon rains be affected?
- What data has been reported about the composition of the emissions of the oil well fires? Where is it going? What are the reported effects?
- What local problems have been reported?
- What are some unexpected reported properties of the soot?

SOURCES

Science News, Feb. 2, 1991, p. 71.

Science News, vol 140, 1991, p. 24.

Nature, May 30, 1991

Science, June 14, 1991
C. Veldon et al.

SYMBOL KEY	
●	Topic of average difficulty
Δ	Long-term assignment
#	Project requires special facilities, equipment or supplies
o	Large public or college library required
∗	Safety precautions required
+	Highly Technical; specialized knowledge required

STRATEGIES FOR DEVELOPING A SUSTAINABLE AGRICULTURE ☉

In order to provide for the growing human population, food production must expand; but it must do so using strategies which minimize soil erosion, desertification, salinization of the soil, pollution of groundwater, and other environmental damage associated with farming. Some agricultural research is aimed at biologically engineering plants such as corn that will "fix" nitrogen and thus decrease the need for the inorganic fertilizers that pollute water and degrade the soil. "Water harvesting" is an ancient technique which will save irrigation water: the land is shaped so that water will run down from large upland areas into collectors or into smaller areas containing crops. Trickle irrigation using gravity decreases water usage along with salinity. Techniques have also been developed to recycle wastewater so as to minimize chemical pollution of the environment. "Multiple cropping" uses crop rotations along with intercropping (other crops or trees share a field). Intercropping can enrich the soil while decreasing soil erosion, pest populations, and weeds. Crops can be selected so that an insect that thrives on one of the plants will be controlled by its predator which resides on another plant sharing the same field. Such a strategy is part of integrated pest management (IPM). IPM includes many other techniques such as chemical and biological pest control along with physical manipulation of the soil.

IDEAS TO EXPLORE

- What have been some of the harmful effects of agriculture on the environment?
- What are some of the important strategies of "sustainable agriculture" and how do these strategies reduce damage to ecosystems?
- Compare practices in both the developed and developing nations.
- What are some directions presently being taken by researchers to develop "sustainable agriculture"?

PROJECT

Environmental Impact of Various Farming Methods [•]

Select a specific farm, or farm community in your state and determine present methods of pest management, fertilization, and water usage. Assess environmental impacts of these methods, and based on your literature research suggest changes that will be beneficial to both farmer and environment.

SOURCES

Scientific American, Sept. 1989, p. 128.
P. R. Crosson and N. J. Rosenberg

Consultative Group on International Agricultural Research (CIGAR), World Bank, Washington, D.C.

Science, vol 217, p. 215.
D. L. Plucknett and N. J. H. Smith

7

HEALTH, NUTRITION, AND FITNESS

CAROTENOIDS AS PROTECTION AGAINST CANCER ● ○

Carotenoids are a class of compounds related to vitamin A which have yellow-to-red colors. They occur mostly in green and yellow vegetables but are also present in egg yolks and even shark oil. Beta-carotene and canthaxanthin are two carotenoids that are being studied because of their suspected role in suppressing the development of some cancers. Researchers at a Japanese medical school observed a significant suppression in the proliferation of human neuroblastoma cells after adding alpha- or beta-carotene. Experiments showed that the carotenoids suppressed the expression of a proto-oncogene that codes for cell-growth enhancing proteins.

IDEAS TO EXPLORE

- What data exists that supports the cancer-protecting effects of carotenoids in the diet?
- How is retinol, the primary form of vitamin A, related to risk of some cancers? What mechanisms have been proposed to explain proposed effects?
- What are oncogenes? proto-oncogenes?
- What other experiments should be performed to determine if some carotenoids decrease the risk of some cancers?

PROJECT

Carotenes and Cancer •

Examine the diets of different cultures, determining sources of carotenes. Compare with rates of different cancers in those cultures. Members of some nonindustrialized cultures (China, for example) may be more helpful in establishing the role of carotenoids in cancer risk because they are not subjected to some other risk factors such as air and water pollution, alcohol consumption, high meat diets, and radiation.

SOURCES

Journal of the National Cancer Institute, Nov. 1, 1989
M. Murakoshi et al.

American Journal of Epidemiology, March, 1990
S. Graham et al.
State University of New York at Buffalo

World Health Organization

SYMBOL KEY	
•	Topic of average difficulty
Δ	Long-term assignment
#	Project requires special facilities, equipment or supplies
○	Large public or college library required
*	Safety precautions required
+	Highly Technical; specialized knowledge required

THE EFFECT OF DIET ON THE IMMUNE RESPONSE ●○

Nutrition influences an individual's resistance to disease by strengthening the immune system. Good nutrition also promotes recovery from disease. Some specific effects of nutrients on the immune system include the following:

(1) Protein deficiency can inhibit T-lymphocyte and antibody synthesis.

(2) Macrophages with high levels of cholesterol exhibit a depressed phagocytic ability.

(3) Impairment of the cell-meditated immunity and reduced ability of white cells to function have been shown in some obese people.

(4) Zinc deficiency can reduce the size of the thymus and decrease the number of mature T cells as well as the T helper to suppressor ratio.

(5) Folic acid deficiencies depress T-cell numbers while vitamin B_2 and B_1 deficiencies lower B-cell activities. Vitamin C appears to play a major role in the body's protection against infection and diseases. For example, cells capable of phagocytosis need vitamin C. Vitamin C has been shown to participate in the generation of energy needed for the secretion of immunoglobulins, interferons, and other cytokines from cells.

IDEAS TO EXPLORE

- What biochemical processes are involved in the immune response?
- How do nutrients affect the immune response?
- What special role does vitamin C play in the body's defenses against infection and disease? Have studies shown a role of vitamin C in cancer prevention?

SOURCES

Food Technology, vol 41, p. 112.
A. Libldings

RN, vol 53, p. 67.
D. Cerrato

Nutrition and Immunology, vol 252, p. 1443.
R. Chandra

"Human Nutrition and Dietics," 1986
Davidson and Passmore
Churchill Livingstone, NY

SYMBOL KEY	
●	Topic of average difficulty
Δ	Long-term assignment
#	Project requires special facilities, equipment or supplies
○	Large public or college library required
*	Safety precautions required
+	Highly Technical; specialized knowledge required

FIBER IN THE DIET ☐

Fiber has been associated with lowering the risk of heart disease and some cancers. Fibers are complex undigestible substances present in food products from both animals and plants. They have a variety of properties; for example, some, such as cellulose, are insoluble in water while others—like oat bran and pectin—are water soluble. Their different properties result in different interactions in the body and therefore their health-effects vary. Pectin and oat bran fibers are thought to bind bile acids which are formed from the breakdown of cholesterol by the liver. Binding the bile acids allows the liver to break down more cholesterol, thus lowering blood cholesterol levels. Water-insoluble fibers may reduce the risk of colon cancer by providing roughage, thus hindering constipation. A combination of fibers has recently been proposed to reduce the risk of breast cancer.

IDEAS TO EXPLORE

- What are the different kinds of fibers in the diet and in which foods are they present?
- What effects are they thought to have on the risk of heart disease and some cancers?
- What are some proposed mechanisms for their effects?
- The National Research Council emphasizes a diet that provides a good mixture of soluble and insoluble fibers. Why?

PROJECT

Comparison of Fiber Content in Various Diets •

Determine the fiber content of your diet and the diets of your friends. Compare with your parents and your friends' parents. How much of water-insoluble and water-soluble fibers are present? Do family members or members of the same generation ingest similar amounts and kinds of fibers?

What might your result suggest about the risk of heart disease and some cancers? Based on fiber alone, are risks more similar in the same family or in the same generation?

SOURCES

"Dietary Fiber: Chemistry, Physiology, and Health Effects"
Davi Kritchevsky, et al.
George Vahouny Fiber Conference, 1988
Washington, D.C.

Handbook of Dietary Fiber in Human Nutrition, 1986
Gene A. Spiller
CRC Press
Boca Raton, Fla.

"Unconventional Sources of Dietary Fiber: Physiological and In Vitro Functional Properties," 1983
Ivan Furda
American Chemical Society
Washington, D.C.

DENTAL HEALTH AND SUGAR CONSUMPTION •

This topic may be of special interest to future dental health workers or to those who are particularly prone to tooth decay (caries). Tooth enamel consists mostly of hydroxyapatite, $Ca_{10}(PO_4)_6(OH)_2$, which will dissolve under acidic conditions forming caries:

$$Ca_{10}(PO_4)_6(OH)_2 \text{ (solid)} \longrightarrow 10Ca^{2+} + 6PO_4^{3-} + 2OH^-$$

The pH of saliva is normally neutral at about 6.8 so that little hydroxyapatite dissolves. In addition, saliva contains buffers which help neutralize the acid produced by plaque, a gelatinous mass consisting mostly of bacteria. Experiments have shown that after rinsing with a glucose mouthwash, oral pH drops to 5.5 within two minutes and remains at that level or below for about 20 minutes. Cavity formation occurs at a pH of 5.5 or below. The pH returns to normal values after about 40 minutes.

SYMBOL KEY	
•	Topic of average difficulty
Δ	Long-term assignment
#	Project requires special facilities, equipment or supplies
o	Large public or college library required
*	Safety precautions required
+	Highly Technical; specialized knowledge required

PROJECT

Measuring Oral pH Changes •

Ask each participant in your study to rinse with a series of glucose solutions with varying concentrations. Collect samples of saliva before and after rinsing and measure the pH of each sample. Your results will determine whether the concentration of glucose influences oral pH. Is there a threshhold concentration below which there is no significant drop in pH?

Repeat this experiment using other sugars such as sucrose or fructose to determine whether they exert a similar or different effect on oral pH. Look for individual differences among your subjects. Note whether family members tend to respond more similarly to sugar than non-family members. Notice any similarities in oral pH response to sugar between members of the same age group.

The pH can be measured with pH paper. The color of the strip indicates the pH of the solution. (See Resources for the name and address of a chemical supply company where you can buy pH paper.)

SOURCES

Chemical and Engineering News

HEALTH EFFECTS OF EXPOSURE TO LOW LEVEL RADIATION • Δ

Low level radiation is generally defined as radiation which does not cause immediate symptoms of radiation sickness but is suspected of having long term harmful effects such as cancer and leukemia. Sources of radiation are both natural and artificial. Artificial sources include radioactive isotopes emanating from a nuclear facility used to produce electric power or to generate plutonium for nuclear weapons, medical procedures such as X-rays or radioactive dyes used for treatment of tumors or for diagnostic purposes, radiation from computer monitors and microwave ovens, and radiation from high tension wires. Fallout associated with above ground testing of atomic weapons and from malfunctioning nuclear power plants is another source of exposure to low level radiation. Natural sources include exposure to soil rich in uranium. Such an area would also contain high concentrations of radioactive isotopes in drinking water and therefore, most likely in milk and milk products as well as in vegetables and meat products produced in the area.

PROJECT

Radiation Exposure in Your Locality •

Determine the various sources of radiation exposure for you and/or members of your family or community throughout your or their lives. Estimate the amount of each exposure. Call your physician to inquire about radiation from chest X-rays and your dentist to ask about dental X-rays. Compare your total exposure with maximum amounts recommended by the National Institute of Health. Compare your total or yearly exposure with that of your parents, classmates, or other members of your community.

SYMBOL KEY

- • Topic of average difficulty
- Δ Long-term assignment
- # Project requires special facilities, equipment or supplies
- ○ Large public or college library required
- ∗ Safety precautions required
- + Highly Technical; specialized knowledge required

THE EFFECTS OF AIR POLLUTION ON HEALTH ⊡

The deleterious effects of air pollution became apparent during the acute rise of respiratory illness and deaths from air pollution in Donora, Pennsylvania in 1948 and in London in 1952. In the United States today more than half of all Americans live in areas that do not even meet national air quality standards designed to protect health. Children and adults with chronic respiratory and heart disease represent populations that are most at risk, especially when exercising. A recent report by the American Lung Association estimates that health benefits of $50 billion could be gained annually if air quality standards were met.

IDEAS TO EXPLORE

- Which pollutants are associated with harmful effects on health? What studies show these effects?
- What are the existing national air quality standards? How were these concentrations determined? Does your state have more stringent standards?

SYMBOL KEY	
•	Topic of average difficulty
Δ	Long-term assignment
#	Project requires special facilities, equipment or supplies
○	Large public or college library required
*	Safety precautions required
+	Highly Technical; specialized knowledge required

PROJECT

Air Pollution in the Workplace •

Measure air pollutants associated with different occupations to assess whether some workers are exposed to elevated levels. For example, measure carbon monoxide (CO) in an enclosed parking garage, or in a commerical kitchen using a gas stove. Many air pollutants can be measured using color detector tubes with grab sampling. The detector pump and color tubes, along with instructions, are available from SKC Inc. (334 Valley View Road, Eighty Four, PA 15330-9614). Be sure to make several measurements at different times of the day. Also, take two or three measurements during each sampling period to decrease experimental error associated with your results.

SOURCES

New England Journal of Medicine, vol 321, 1989, p. 1426. E. Allred

"Cancer Risk from Outdoor Exposure to Air Toxics," U.S. EPA

Office of Air Quality Planning and Standards: September 1989, p. ES-2.
Research Triangle Park, NC

SOME PHYSIOLOGICAL RESPONSES TO STRESS

Respiration and heart rates along with blood pressure tend to increase after a stressful event due to a central nervous system reaction involving the release of epinephrine.

IDEAS TO EXPLORE

- What is stress?
- How does the body respond to stress?
- What are physiological responses to long term stressful situations?
- How can the harmful effects of stress be decreased?

PROJECT

Responses to Violence 

Show several volunteers violent scenes from a movie and then measure respiration, heart rate, and blood pressure every 5 minutes or less until these vital signs return to normal. Perform this experiment on three subsequent days to determine if the subjects become less affected by the violence. You can use two groups of volunteers to determine if reactions to stress are affected by the time of day (mornings vs nighttime for example) during which the stress occurs. Try to use subjects of similar age. You could also perform the experiment on a group of men and a group of women to see if there are differences in their response to violence.

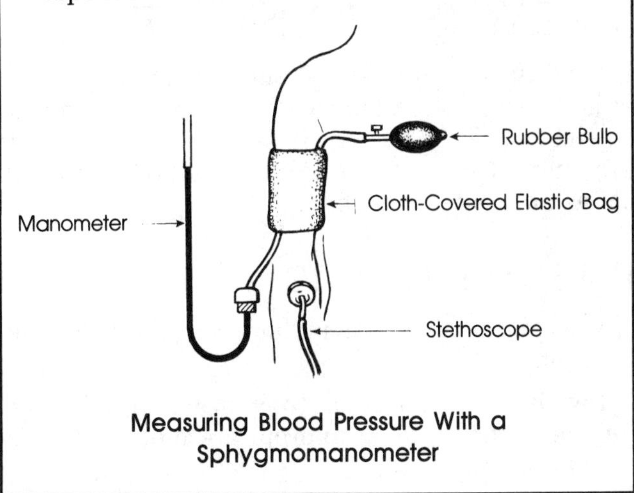

Measuring Blood Pressure With a Sphygmomanometer

SOURCES

A general physiology textbook will discuss the body's response to stress in detail. An example is
Principles of Anatomy and Physiology, 1984
G. Tortora and N. Anagnostakos
Harper and Row, New York

SYMBOL KEY	
●	Topic of average difficulty
Δ	Long-term assignment
#	Project requires special facilities, equipment or supplies
o	Large public or college library required
∗	Safety precautions required
+	Highly Technical; specialized knowledge required

PROLONGED EXERCISE AND ENERGY METABOLISM • ○

Carbohydrates alone cannot support prolonged exercise; lipid components appear to be more important for extensive physical exercise. In fact, the higher the level of training, the greater the athlete's ability to utilize lipids as an energy source.

IDEAS TO EXPLORE

- What biochemical data is there to support the above statements?
- How is the data used to improve training techniques and select an appropriate diet for the individual athlete?
- Could biochemical data be helpful in identifying a future Olympian?

SOURCES

Metabolism, vol. 29, 1980, p. 662.
P.D. Thompson et al.

Journal of Applied Physiology, vol. 28, 1970, p. 251.
D.J. Costill

PROJECT

The Effect of Jogging, Swimming, or Bicycling on the Cardiovascular System •

Ask several of your friends who are presently not swimming, jogging, or bicycling on a regular basis to participate in your study. Include yourself if you qualify. The idea is to measure the effect of aerobic exercise on resting pulse rate and blood pressure. You can also measure the time it takes for pulse rate to return to its resting value (recovery time) after exercising. All participants should be checked by their physician prior to participating in the study. As with all exercise programs, start slowly and increase duration of exercise gradually at your own pace. Measure the resting pulse rate and blood pressure of all participants at the beginning of the study and once a week thereafter. Ask a health care professional for instructions on pulse rate and blood pressure measurement. At some time during the study, start to measure the time it takes for pulse rate to return to its resting value after exercise. Take the pulse rate after each five minute interval.

The questions which you might study, depending on the number of subjects, include:

- Do the cardiovascular systems of females and males respond similarly to aerobic exercise?

- Does the degree of cardiovascular response depend on the fitness of the subject at the beginning of the exercise program?
- Do the three activities have similar effects on the cardiovascular system?

SOURCES

Consult a physiology textbook. An example is *Principles of Anatomy and Physiology*, G. Tortora and N. Anagnostakos, Harper and Row

SYMBOL KEY	
●	Topic of average difficulty
Δ	Long-term assignment
#	Project requires special facilities, equipment or supplies
○	Large public or college library required
∗	Safety precautions required
+	Highly Technical; specialized knowledge required

SIDE-STREAM TOBACCO SMOKE ● ○

It has been recently recognized that smoking can be harmful to the health of nonsmokers as well as smokers. Nonsmokers are exposed to tobacco smoke emanating directly from the lit tip of the cigarette, cigar or pipe, as well as to the smoke exhaled by the smoker. The two kinds of smoke are chemically different because exhaled cigarette smoke, for example, has been filtered somewhat by the length of the remaining cigarette, the cigarette's filter, as well as by the respiratory system of the smoker.

IDEAS TO EXPLORE

- What recent studies have indicated that side-stream tobacco smoke is harmful?
- Does side-stream smoke increase your risk of lung cancer? heart disease?
- Is the smoke emanating from a lit cigarette more carcinogenic than the smoke inhaled by the smoker?

SOURCES

Annals of the New York Academy of Science 1989, vol 562, p. 74.
D. Rush et al.

New England Journal of Medicine
Dec 1988, vol 319, p. 1452.
J. Fielding et al.

8

MEDICINE

LOW-FREQUENCY ELECTROMAGNETIC FIELDS AND CANCER ⊡

Recent epidemiologic studies link cancers with exposures to low-frequency electromagnetic fields (ELF). These fields are generated by the flow of electricity and therefore emanate from household appliances as well as from transmission lines. Wertheimer and Leeper first reported an association between electric power lines and childhood leukemia. After reviewing the available scientific literature, the EPA acknowledged that ELF may be a human carcinogen. Recent studies also showed a link between electric blankets and risk of brain tumors in children whose mothers used them during the first trimester of pregnancy. At Yale University an experiment is being conducted in which a large cohort of pregnant women is being followed to determine the effects of ELF exposures on their children.

IDEAS TO EXPLORE

- What are the sources of electromagnetic fields that people are exposed to?
- What data is there to suggest a causal link between ELF and human cancers? What data is lacking?
- What is known about how ELF can cause cells to become malignant?
- Are there occupational exposures to ELF?
- Can you suggest studies that could help determine if ELF causes cancer?

PROJECT

ELF Exposure in Your Household [• #]

Determine sources of ELF in your household. Find out or measure the strength of these fields and the time of exposure of yourself and other family members. Compare the exposure per year for the different members of your family. How can these exposures be reduced? How do these exposures compare to those studied in the scientific literature?

SOURCES

Science News, Apr. 21, 1979
N. Wertheimer and E. Leeper

Science News, June 30, 1990, p. 404.

U.S. EPA Health and Environmental Assessment Office directed by William Farland

American Journal of Epidemiology, May, 1990.
David Savitz, University of North Carolina at Chapel Hill, NC

PROJECT

The Effect of Cigarette Smoking on the Sense of Smell ⊡

Some studies have shown a decline in olfactory functioning (sense of smell) as a result of cigarette smoking while other studies do not. Your project can add to the current data.

Select two groups of subjects: one group will consist of past smokers while the other group will consist of nonsmokers with no smoking history. Try to find participants who are approximately the same age. (Gender may also be important.) Prepare solutions of an odorant (a volatile liquid) that have different concentrations. (The greater the concentration the greater the perceived odor.) It is probably a good idea to select an odorant which is pleasant to help keep your subjects in the study. After your subject smells the vapor above the first solution presented, ask the subject to indicate his or her perceived odor intensity at the center of a line graph: ├┼┼┼┤. The subject should then indicate the perceived intensity of all other samples relative to the first. It is always a good idea, when possible, to conduct a double-blinded study. Therefore, after you prepare the solutions cover the labels so that you will not know the concentrations of the samples as the subjects proceed with the experiment.

SOURCES

Odor studies are frequently performed by many industries before marketing certain new products. Thus the protocol for odor studies appears in many journals including *Science*, *Nature*, and *Scientific American*.

Many books have also been published on perceived odors including *Human Responses to Environmental Odors*, 1974, edited by Amos Turk, James Johnston Jr, and David Moulton, Academic Press, Inc., New York, NY

SYMBOL KEY	
●	Topic of average difficulty
Δ	Long-term assignment
#	Project requires special facilities, equipment or supplies
o	Large public or college library required
*	Safety precautions required
+	Highly Technical; specialized knowledge required

ALZHEIMER'S DISEASE • Δ ○

Alzheimer's disease poses a major health problem for the elderly. Recent estimates indicate that almost half of the U.S. population 85 years and older will develop this disorder. The cause of the disease is unknown, the diagnosis is difficult, and in addition, Alzheimer drugs offer limited improvements to the patient.

- *Causes:* The symptoms of Alzheimer's disease are associated with the death of nerve cells that secrete acetylcholine, a neurotransmitter in the brain. Proteins called amyloid plaques are known to accumulate in the brain and are seen only upon autopsy. Various causes have been suggested such as mitochondrial mutations, viral infections, and elevated aluminum ion ingestion.
- *Diagnosis:* Alzheimer's disease can only be positively diagnosed upon autopsy by the presence of amyloid plaques in brain tissue. Suspected Alzheimer patients exhibit a progressive dementia including significant memory loss. Other disorders associated with aging, such as senility, show similar symptoms making diagnosis of the disease difficult.
- *Cures:* In order to cure or prevent a disease its cause must generally be known. Recent attempts at developing drugs focus on decreasing acetylcholine breakdown. Research, however, implies that other neurotransmitter systems may be involved so that these drugs may be of limited value. Another direction of research involves developing drugs that can decrease oxidative damage that disrupts mitochondrial metabolism.

IDEAS TO EXPLORE

- What are the various causes of Alzheimer's disease that have been proposed by the scientific community? What data supports each of these causes?
- How is Alzheimer's disease diagnosed?
- How can it be differentiated from other psychiatric disorders such as senile dementia?
- How are Alzheimer patients managed? What drugs have been developed? Are they effective?

SOURCES

Science News, vol. 137, No. 8, p. 120.
R. Weiss

Many articles in *New England Journal of Medicine*, *Science*, *Nature*, and *Scientific American*.

SYMBOL KEY	
●	Topic of average difficulty
Δ	Long-term assignment
#	Project requires special facilities, equipment or supplies
o	Large public or college library required
∗	Safety precautions required
+	Highly Technical; specialized knowledge required

ALUMINUM AND ALZHEIMER'S DISEASE ● ○

Elevated concentrations of aluminum have been detected in four regions of the brain of Alzheimer patients. It appears in senile plaques and in the tangle-bearing neurons characteristic of the disease. Aluminum has also been implicated in parkinsonian dementia, dialysis dementia, and in some forms of epilepsy.

Currently, there is much debate regarding aluminum intake and the risk of Alzheimer's disease. Martyn et al. showed a strong correlation between the risk of the disease and aluminum concentrations in drinking water in an epidemiological study described in *The Lancet*.

IDEAS TO EXPLORE

- What data suggests that aluminum plays a role in developing Alzheimer's disease?
- What theories have been proposed to explain how aluminum permeates the blood-brain barrier?
- Give the results of some studies scientists believe do not support the aluminum hypothesis.
- What is the difference between aluminum intake and aluminum uptake?
- Aluminum occurs naturally in foods because it is a substantial component of the earth's crust. It is also present in food additives, medications such as antacids and analgesics, deodorants, foods prepared in aluminum cookware, drinking water treated with alum, a flocculent, in tea and in coffee made from some automatic coffeemakers (Brisk et al.). Where else does it occur?

- A new drug, desferal, is purported to slow the progress of Alzheimer's disease. Discuss the study. (The researchers of the study attribute the drug's claimed efficacy to its aluminum sequestering capacity thus removing aluminum from the body.) What further studies are needed to support or contradict these claims?

PROJECTS

1. Determining Aluminum Intake • ○ # *

Determine sources of aluminum and approximate your intake per week. Compare with that of your parents and grandparents at your age. You will need to consult the literature for dietary sources and associated concentrations of aluminum in order to estimate intake. If you suspect the presence of aluminum in other dietary sources (for example, soda from aluminum cans), you can measure these concentrations using chemical test kits such as those available from EM Science. Is aluminum in the same chemical form in soda as in vegetables or meat, for example?

2. Factors Affecting Amount of Aluminum in Foods • ○ # *

For a more complete project on exposure to aluminum you might determine if more aluminum leaches from the surface of an aluminum can when left open and therefore exposed to air? when fluoride ions are present? at lower pH values?

SOURCES

The Lancet, vol 1, 1989, p. 59.
C. Martyn et al.

Nature, vol 325, 1987, p. 202.
K. Tennakane and S. Wickramanayake

Science, vol 208, 1980, p. 297.
D. D. Perl et al.

Easy-to-use chemical test kits for aluminum, lead and other ions are sold by EM Science, P.O. Box 70
480 Democrat Road, Gibbstown, NJ 08027

SYMBOL KEY

- • Topic of average difficulty
- Δ Long-term assignment
- # Project requires special facilities, equipment or supplies
- ○ Large public or college library required
- * Safety precautions required
- + Highly Technical; specialized knowledge required

AIDS: THE TRAGEDY OF THE TWENTIETH CENTURY • Δ

The World Health Organization has recently predicted that by the year 2000 *forty million* people may be infected by the AIDS virus called HTLV-III. AIDS, or Acquired Immunodeficiency Syndrome, is caused by a retrovirus that invades the DNA of T-lymphocytes compromising the immunity of the victim against a variety of microbes. AIDS is contagious and is one of the leading causes of death among young people in the United States. Many laboratories in the world are studying AIDS and working to develop drugs to be used in the management of AIDS as well as a vaccine to prevent the disease and spread of the virus.

IDEAS TO EXPLORE

- Which nations have the highest incidences of AIDS? What are the primary modes of transmission in each of these areas?
- Is the retrovirus causing AIDS the same in each region?
- What treatments are prescribed in each country?
- What attempts are being made to limit the spread of infection? How successful are they?
- What approaches are being taken to develop a vaccine for AIDS?

The Changing Epidemology of AIDS ● ○

AIDS is rising sharply among American women, in particular, poor black and Hispanic women. WHO predicts that by the year 2000 the number of new cases among women worldwide will start to equal the number of newly diagnosed men. Although heterosexual intercourse is the predominant mode of transmission in most countries, it now accounts for only a small percentage of AIDS cases in the United States. Heterosexual transmission in particular is on the rise in America.

IDEAS TO EXPLORE

In the United States the groups associated with a high risk of HIV infection include intravenous drug users, homosexual men, and hemophiliacs.

- In what groups is the incidence of AIDS increasing?
- Why is AIDS rising among American women?

How Does the AIDS Virus Incapacitate the Immune System ● ○

The processes involved in the incapacitation of the immune system are being elucidated by many research laboratories. Such knowledge may one day lead to a cure or vaccine as well as shed light on cancer and other diseases.

IDEAS TO EXPLORE

- How does the AIDS retrovirus affect the immune system?
- What other organ systems are directly affected?

Common Opportunistic Infections of AIDS Patients

Because the immune system of AIDS patients is compromised, opportunistic infections such as pneumocystic carinii, toxoplasomosis, and cryptococcus affect and often kill the patient.

IDEAS TO EXPLORE

- What are the major opportunistic infections of AIDS patients?
- What other patients are prone to these diseases? Why?
- Can they be prevented? Is AZT effective in their prevention?

Neurological Effects of AIDS

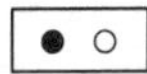

Because the AIDS virus can cross the blood-brain barrier, patients often suffer from psychiatric disorders including dementia.

IDEAS TO EXPLORE

- What are the major neuropsychiatric disorders of AIDS patients?
- What are their causes?
- How are they treated?

Is An AIDS Vaccine Possible? 

Jonas Salk, who developed the polio vaccine, is using a similar approach to finding an AIDS vaccine.

IDEAS TO EXPLORE

- What are the results of Salk's preliminary experiments with an AIDS vaccine?
- What other laboratories are working on an AIDS vaccine? What approach or approaches have they taken?

SOURCES

Many journals contain articles on AIDS including *Science*, *Nature*, *Scientific American*, and medical journals such as the *New England Journal of Medicine*, *Journal of the American Medical Association*, and *The Lancet*.

SYMBOL KEY	
•	Topic of average difficulty
Δ	Long-term assignment
#	Project requires special facilities, equipment or supplies
○	Large public or college library required
*	Safety precautions required
+	Highly Technical; specialized knowledge required

THE EFFECT OF LOW BODY TEMPERATURE ON NEURONS ● ○

Heart attacks and strokes decrease the amount of oxygenated blood in the brain causing damage to neurons. Studies with gerbils have shown that brain-neuron destruction is reduced by lowering body temperature (hypothermia). Hypothermia has also been used to explain the survival of children who have been submerged in cold water for prolonged periods without suffering brain damage. Such data implies that lowering body temperature may be useful as an emergency procedure for victims of heart attacks or strokes.

IDEAS TO EXPLORE

- What is known about the effects of low temperatures on brain tissue?
- How does hypothermia explain the survival of oxygen-deprived patients?
- Which patients might benefit from a decreased body temperature?

SOURCES

Science News, vol 137, No. 10, p. 154.
R. Weiss

Journal of Neuroscience, Jan. 1990.
A. Buchan and W. Pulsinelli
Cornell University Medical College in NYC

Science, Jan. 12, 1990.

THEORIES OF AGING • Δ

It is well known that the body undergoes a decline with aging. These changes include hair loss, decrease in hearing, sagging skin, memory loss, and increase in rates of heart disease and cancer. Two theories proposed are the damage theory and the program theory. The damage theory attributes changes in cells associated with aging to destructive environmental stresses such as UV radiation. The program theory attributes aging to a biological clock that exists within the DNA of cells. A longitudinal study is currently being conducted on aging at Johns Hopkins University by Jerome Fleg, M.D. to determine how much of aging is due to the sedentary lifestyle often adopted by the elderly. The study is supported by the National Institute on Aging.

IDEAS TO EXPLORE

- What are the various theories on aging? What data are used to support each theory?
- What mechanisms have been proposed to explain increased cancer risk with aging?
- Do people age differently in different cultures?

SOURCES

Many articles have appeared in *Science*, *Nature*, *Scientific American*, and in medical journals such as the *New England Journal of Medicine* and *Journal of the American Medical Association* (JAMA). Also *American Family Physician*, Sept. 1989, vol. 40, p. 195, S. Goldstein et al.

EFFECT OF WAR ON THE INCIDENCE OF HEART DISEASE

A correlation has been made between exposure to wartime events and the risk of coronary heart disease for civilians. Patients with coronary heart disease in a hospital in Beirut were interviewed and filled out extensive questionnaires to establish exposure to wartime events and to known coronary risk factors. These patients were then compared to a similar cohort randomly selected from the hospital visitors. After corrections for the known risk factors were made, it was determined that exposure to wartime events increased the chances of heart disease. The risk also rose as the frequency of exposure increased so that a person subjected to four events or more was eleven times more likely to develop the disease than someone exposed to only one event.

IDEAS TO EXPLORE

- What are the limitations of this study?
- Has an increase in heart disease been shown for victims of natural disasters?
- What longtime effects could exposure to war or disasters cause to account for the increased risk of heart disease?

SOURCES

American Journal of Epidemiology, Oct. 1989.
A. Sibai and H. Armenian

BREAST CANCER: THE SILENT EPIDEMIC ● ○

Approximately 44,000 American women die annually from breast cancer, more than the number of Americans who succumb to AIDS. It is predicted that one out of nine American women will eventually die of this disease. Chinese and Japanese women have a much lower incidence of breast cancer in their native countries but the incidence rises dramatically for Chinese and Japanese women who live in the U.S.

Many studies of European and North American women have indicated that a high fat diet is a risk factor of breast cancer. However, a recent study of Chinese women shows a relationship between risk of breast cancer and dietary fat in low fat diets as well.

IDEAS TO EXPLORE

- What is the incidence of breast cancer in other industrialized countries? in nonindustrialized countries?
- What are known or suspected risk factors?
- What is known about dietary factors? radiation exposure? exposure to carcinogens?
- Are there elevated rates of breast cancer associated with some occupations?
- Could soy-rich foods help prevent breast cancer?

SOURCES

Cancer Research, Aug. 15, 1990.

Science News, May 12, 1990, p. 296

National Cancer Institute

Centers for Disease Control

World Health Organization

SYMBOL KEY	
●	Topic of average difficulty
Δ	Long-term assignment
#	Project requires special facilities, equipment or supplies
○	Large public or college library required
∗	Safety precautions required
+	Highly Technical; specialized knowledge required

EFFECTS OF LEAD EXPOSURE ON MENTAL DEVELOPMENT OF CHILDREN ● ○

It is well known that exposure to high doses of lead is harmful to both adults and children. Symptoms include cognitive deficits in children and progressive renal disease in adults. Recently, however, it has been shown that exposure to low doses of lead also retards the mental development of children. As a result the Centers for Disease Control lowered the acceptable blood level of lead for children. Urban children are especially at risk due to inhalation of fumes from combustion of leaded gasoline and ingestion of paint chips containing lead. Fetal lead exposure has been shown to affect the mental development of infants. Low doses are ingested from drinking water flowing through pipes with lead containing solder and from contamination of acidic foods and beverages in metal cans with lead containing solder.

IDEAS TO EXPLORE

- What are the effects on children of exposure to high does of lead?
- What are these "high dose" lead sources?
- What children are especially at risk?
- What is now known about "low dose" lead sources and their effects? What children are at risk?
- What are the mechanisms of neurotoxicity and their relationship to behavorial changes?

PROJECT

Lead Exposure of Family and Friends • # *

Identify sources of lead exposure for you and members of your family. Compare with sources of lead exposure for your friends and their families. Are differences greater between generations or between families? Measure the concentrations of lead in your drinking water and in acidic beverages from cans (cola drinks, fruit juices, tomato juice) with an ion-sensitive electrode or by using a lead test kit that can be purchased at a chemical supply company.

SOURCES

Merck Manual (Lead Poisoning)

Environmental Health Perspective, Nov. 1987, p. 59.
T.W. Clarkson

Pediatrics, Nov. 1987, p. 721

The New England Journal of Medicine
Vol 316, 1987, p. 1037
D. Bellinger et al.

Lead test kits are available at EM Science,
P.O. Box 70, 480 Democrat Rd, Gibbstown, NJ 08027

THE EFFECTS OF COCAINE ON FETAL DEVELOPMENT

Drug addiction is clearly a major problem in our society which has affected even newborns. Chronic drug abusers often give birth to newborns who have low birthweight, abnormal length and head circumference, and are addicted themselves. These newborns often exhibit neurobehavioral deficiencies, malformations of genito-urinary tract, as well as impaired orientation, motor abilities, and reflexes.

IDEAS TO EXPLORE

- What are some known effects of drug abuse on the fetus?
- What is known about how these drugs cause the effects?
- What treatments are used?
- What is the prognosis for these newborns?

SOURCES

Obstetrics and Gynecology, Dec. 1989.

Pediatrics, Oct. 1990, p. 639.

THE NEUROBIOLOGY OF SLEEP AND DREAMING ● ○

One of the universal characteristics of all life, including unicellular organisms, is a rhythm of rest and activity. In humans it is a complex function and consists of a cycle of dreamless Non-Rapid Eye Movement (NREM) sleep followed by Rapid Eye Movement (REM) sleep with dreaming. Evidence from studies of animals and of human sleep disorders suggests that sleep is necessary for energy regulation as well as for information processing by the brain.

IDEAS TO EXPLORE

- What is known about sleep and body temperature regulation? immune system functions? cognitive functions?
- What are the effects of sleep deprivation on concentration? on motivation? on memory?
- What are some known sleep disorders and what information from their study has contributed to the neurobiology of sleep?
- What are the present theories about sleep?

PROJECT

Effect of Sleep on Concentration and Memory ⊡

Select a group of your friends to participate in a study relating sleep and ability to concentrate and memorize. First, ask each participant to record the number of hours slept the night before. Then, ask each one to memorize a list of numbers, words, or phrases and to perform some arithmetic calculations. Record both the time each participant requires to complete the calculations and the accuracy of each one's results. Repeat this procedure at the same time of day over a period of several weeks so that your results will include days or periods of sleep deprivation. Is there a relationship between sleep and performance on your tests? You could also follow the same procedure with members of your parents' generation and determine whether age makes one more or less dependent on sleep. The references given below may be helpful.

SOURCES

British Journal of Psychology, 1989, vol 80, p. 145.
M. Mikulincer et al.

The Journal of Neuroscience, Feb. 1990, vol 10, p. 371
J. Hobson
Department of Psychiatry
Harvard Medical School
Boston, MA 02115

HEALTH EFFECTS OF LONG-TERM EXPOSURE TO COMPUTER SCREENS ⊡

In the late 1970's studies began to show that computer operators may have an elevated risk of developing such health problems as miscarriages, birth disorders in children, and cancer because of their exposure to video display terminals (computer screens). Although subsequent studies have yielded contradictory or different results, the fact that low frequency electromagnetic radiation (EMF) is emitted from video display terminals (VDTs) has prompted Sweden to establish strict low-radiation standards for VDTs. Several long-term studies are presently being conducted at the University of California at Berkeley and the Mt. Sinai School of Medicine in New York.

IDEAS TO EXPLORE

- What studies have shown that health hazards may be associated with VDTs?
- What contradictory data exists?
- Are liquid display monitors (LCD) a better choice? monochromatic screens?
- Since most electric appliances emit electromagnetic radiation, why are researchers more concerned about the effects of computer useage in particular?

SOURCES

For some information about VDTs contact Labor Occupational Health Program at

University of California at Berkeley
2521 Channing Way, Berkeley, CA 94720
(415) 642-5507

SYMBOL KEY	
●	Topic of average difficulty
Δ	Long-term assignment
#	Project requires special facilities, equipment or supplies
○	Large public or college library required
∗	Safety precautions required
+	Highly Technical; specialized knowledge required

9

PALEONTOLOGY

THE SUDDEN APPEARANCE OF ANIMAL SKELETONS ▣

For at least 100 million years marine life existed without hard materials such as shells and scales. Animal skeletons then appeared abruptly. Several theories have been suggested to explain the sudden appearance of the animal skeleton. Recent evidence from the Burgess shale fossil deposit in the Canadian Rockies of British Columbia supports the theory that skeletons evolved in response to the arrival of predators. Other theories attribute the rise of skeletons to a rise in oceanic calcium concentrations and in atmospheric oxygen.

IDEAS TO EXPLORE

- How do these theories explain the appearance of the animal skeleton?
- What data is there to support each theory?
- Which theory do you support and why?

SOURCES

Science News, vol 138, No. 8 p. 120.
W. Stolzenburg

Palaios, Dec. 1989
G. Vermeij, University of California, Davis

Nature, June 28, 1990.
C. Morris and J. Peel, Geological Survey of Greenland in Copenhagen, Denmark

DINOSAURS • Δ

Dinosaurs inhabited the earth during most of the Mesozoic era, disappearing some one hundred and fifty million years ago. They thrived for over 100 million years. (Modern man has lived for only one million years.) The gigantic size of the dinosaurs makes them fascinating as well as mysterious. Paleontologists have limited knowledge of the dinosaurs and are even debating whether they are endotherms (warm-blooded) or ectotherms (cold-blooded). Scientists know so little about dinosaurs because all that remains of them is their bones, which at most can be used to reconstruct skeletons, telling us only about their size, shape, and posture. There exist no similar animals today that can be observed to shed light on how the dinosaurs lived.

IDEAS TO EXPLORE

- What data supports the warm-blooded theory? the cold-blooded theory?
- How would dinosaurs spend their days and nights during the Mesozoic era if they were warm-blooded? cold-blooded?
- Was their size an advantage or disadvantage?

PROJECTS

1. Study One Particular Dinosaur [•]

Select one of the approximately three hundred dinosaurs that are thought to have existed and describe what is presently known about it. Does the evidence suggest that it is an ectothermic or endothermic animal? What did it consume? How did it reproduce?

2. How Dinosaurs Became Extinct [•]

Another very controversial topic regarding these interesting animals is how they became extinct. Several theories have been proposed including a decline in temperature due to the collision of a large meteor with the earth. What theories are being discussed? What evidence is being used to support each theory? Can you think of alternate explanations for their demise?

SOURCES

The Sciences, May/June 1987, p. 56.
J. Ostrom

The Dinosaur Heresies
R. Baker
New York: William Morrow and Company, 1987.

THE ORIGIN OF ROME

Although much is known about the Republican and Imperial eras of ancient Rome through historical texts and archeological excavations, the first era, or the Regal period, is largely a mystery. However, as more advances are made in archeology, information regarding this early period is growing.

Until the nineteenth century, knowledge of the beginnings of Rome was extracted from the analyses and interpretations of early historical texts. Archeologists then began to contribute, but the greatest contribution came in the twentieth century when Giacomo Boni probed older and older layers of the earth, unfolding the evolution of ancient Rome and advancing field archeology in the process. In the last decade other major advances were used to help solve the mystery of the origin of ancient Rome. Boni and his predecessors were limited to small areas with few workers, so they could select only some artifacts to study, passing many others by. Archeologists today use fifty or more workers to study a much larger area, recording all possible clues such as fragments of pottery, rock, or clay. Environmental archeology incorporates knowledge from the earth sciences so that the composition of soil or rock can be used to reveal the past. With each new approach being used to study ancient ruins more knowledge is gained and new theories emerge.

IDEAS TO EXPLORE

- Ancient historians believed that Rome began when a pomerium was traced on the Palatine hill sometime around 750 B.C. Did archeology in the nineteenth century led by Giacomo Boni confirm or repute the beliefs of the early historians such as Livy and Dionysius?
- What contribution did Boni make to field archeology?
- How did recent advances in the techniques of archeology help solve the mysteries surrounding early Rome?
- What did the archeologist Andrea Carandivi discover from his excavations?
- Based on the recent findings, what theory or theories are supported? Were the early historians wrong?

SOURCES

The Sciences, July/August 1989, p. 22.
A. Ammerman
Colgate University, Hamilton, NY

SYMBOL KEY	
●	Topic of average difficulty
Δ	Long-term assignment
#	Project requires special facilities, equipment or supplies
○	Large public or college library required
*	Safety precautions required
+	Highly Technical; specialized knowledge required

NEANDERTHALS: WHERE DO THEY FIT IN HUMAN EVOLUTION ⊡

Neanderthal remains were first discovered in 1856 in Germany. Many well-preserved Neanderthal skeletons have since been found. They suggest that these hominids (members of the evolutionary family of modern man) lived in Europe from about 130,000 to 35,000 years ago. More Neanderthal remains were uncovered in the Middle East along with skeletons of slighter hominids who are thought to be ancestors of modern humans. Dating of these remains implies that early modern man coexisted with Neanderthals in the Middle East for as many as 60,000 years. One group of researchers believe that modern man evolved with little genetic input from Neanderthals. Early Homo sapiens flourished while Neanderthals died out. Other investigators interpreted the fossil finds as indication of interbreeding between early man and Neanderthals to eventually give rise to modern Homo sapiens. This theory suggests that modern man evolved as many as one million years ago in several regions of the world.

IDEAS TO EXPLORE

- What is known about Neanderthals?
- What data supports their coexistence with early Homo sapiens?

- What data supports the theory that they did not contribute genetically to the evolution of modern man?
- What data suggests that they interbred with other early hominids?

SOURCES

Science News, vol 139, 1991, p. 360.

SYMBOL KEY	
●	Topic of average difficulty
Δ	Long-term assignment
#	Project requires special facilities, equipment or supplies
○	Large public or college library required
*	Safety precautions required
+	Highly Technical; specialized knowledge required

10

PHYSICS

SUPERCONDUCTORS: WILL THEY CHANGE OUR LIVES? • Δ

Superconductors are materials which exhibit no resistance to electricity. Those developed so far lose their resistance only at very low temperatures. The development of a superconductor that loses its resistance at moderate temperatures would no doubt spur a technological revolution. As a result, many laboratories all over the world are racing to produce the first room temperature superconductor.

Superconductivity was discovered in the early 1900's by H. Kamerlingh Onnes, who noted that the resistance of mercury could not be measured at about 4K. The property represented an enigma since at low temperatures it was thought that electrons do not have enough energy to cause a current. Meissner and Ochserfeld in 1933 discovered another intriguing characteristic of superconductors—levitation—often called the Meissner effect. In 1972, Bardeen, Cooper, and Schreiffer received the Nobel Prize for their work on superconductivity including their quantum mechanical theory (BCS Theory) which they developed to explain superconductivity. BCS theory, however, set a limit for superconductivity at about 40K which means that if their prediction is correct the practical applications of superconductivity are limited. A major breakthrough occurred in the 1980's when researchers began working with metal oxides doped with other metals. In 1987 Chu and Wu surprised the scientific community when they found that a copper oxide compound laced with yttrium exhibited superconductivity above 77K. The record today is 125K and theoreticians continue to refine their explanations of the phenomenon.

IDEAS TO EXPLORE

- What is superconductivity?
- What theories have been proposed to explain the phenomenon?
- What unusual magnetic properties do superconductors exhibit?
- What are some present and possible applications of superconductivity if it can be developed at higher temperatures?
- Will superconductivity change our lives?

PROJECT

The Meissner Effect * # +

Prepare, analyze, and demonstrate the Meissner effect for the superconductor $YBa_2Cu_3O_{8-x}$. This superconductor can be made with Y_20_3, CO_3, and CuO. (See *Journal of Chemical Education*, vol 64, number 10, p. 847.) When the superconductor is cooled with liquid nitrogen (77K) it will exhibit the Meissner effect. A small magnet will levitate above the superconductor if it is positioned properly. Your product, if shown to have this property, can then be analyzed to determine the oxidation states of copper by an iodometric titration.

Cu^{2+} (aq) + $2I^-$ (aq)→CuI + 1/2 I_2(aq) The liberated iodine is then reacted with standard thiosulfate.

$$I_2(aq) + 2S_2O_3^{2-}(aq) \rightarrow 2I^-(aq) + S_4O_6^{2-}(aq)$$

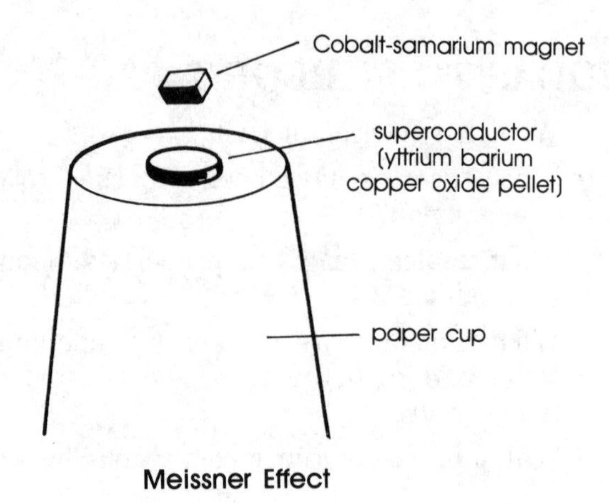

Meissner Effect

SOURCES

Journal of Chemical Education, 1987, vol 64, no. 10, p. 836.
A. Ellis

Physical Review Letters, Mar. 1987.
P. Chu and M. K. Wu
University of Houston
University of Alabama

The Sciences, May/June 1989, p. 44.
M. Schluter, AT&T Bell Labs
Murray Hill, NJ

Superconductivity: The New Alchemy
J. Langone
Contemporary Books, 1989

THEORIES OF SUPERCONDUCTIVITY ○ +

Several theories have been proposed to account for zero resistivity and the unusual magnetic properties of superconductors. These include the London theory, Bardeen-Cooper-Schrieffer theory (BCT) of low temperature superconductivity, and a more recent sudden polarization theory of perovskite structures. Another theory of high-temperature super-conductivity involves the existence of "anyons," a general class of particles.

IDEAS TO EXPLORE

- What theories have been proposed to explain superconductivity?
- How do they explain the Meissner Effect?
- What data supports each theory?
- What are the shortcomings of each theory?

SOURCES

Journal of Chemical Education, vol 64, no. 10, 1987, p. 842
F. A. Matsen

Superfluids,
F. London
Wiley: New York, 1950.

Scientific American, May 1991, p. 58.
F. Wilczek

LIGHTNING ● +

Lightning is a complex phenomenon which can be compared to a charged capacitor undergoing dielectric breakdown, causing an intense flow of electrons. It is the flow of electrons that gives rise to the well-known flash of light.

Cloud-to-ground lightning is the most common kind of lightning in which a separation of charge develops in a cloud. If the resulting electric field is large enough, dielectric breakdown will occur allowing electrons to move. Electrons make 50-meter leaps toward the earth at intervals of one-millionth of a second attracting positive charge in the earth's surface. The subsequent ionization of air molecules near the surface causes dielectric breakdown so that a flash of light is produced.

IDEAS TO EXPLORE

- What are the different types of lightning?
- What is known about the charge distributions that may cause lightning?
- What is a stepped leader? return strobe?
- How can you protect yourself during a lightning storm?

SOURCES

Consult a college physics textbook for a basic understanding of capacitors.

Smithsonian, Aug. 1979, p. 74.
H. Lansford

Ham Radio Horizons, Sept. 1980, p. 12.
D. Logan

Scientific American, Nov. 1988, vol. 259, p. 88.
E.R. Williams

PROJECT

Mapping Electrically Charged Patches [•]

Select a smooth non-conducting surface (such as plastic) on which you can affix adhesive tape. Upon peeling off the tape you will create a charged surface. You can map this surface by dusting differently colored powders such as parsley, sage, flour, Day-Glo powders or others of your choosing onto the charged surface. You will find that some powders work better than others. Some powders will collect onto the negatively charged regions, while others collect in the positively charged sites. These patterns were first reported by Georg Lichtenberg in 1877 and consequently they are called Lichtenberg figures. The Lichtenberg figures you produce will depend on the materials you use and your technique. The patterns can be preserved by affixing several layers of transparent tape over the figure, gently peeling the tape off and then pressing the tape with the dust stuck to it onto a white sheet of paper. Because photocopying machines are based on electrostatics, experiments like these have been done by Xerox Corporation. Details of mapping electrostatic charging with Lichtenberg figures are given in the reference sited below.

SOURCE

Scientific American, Jan 1986, p. 114.
J. Walker

SYMBOL KEY	
●	Topic of average difficulty
Δ	Long-term assignment
#	Project requires special facilities, equipment or supplies
○	Large public or college library required
∗	Safety precautions required
+	Highly Technical; specialized knowledge required

11

PSYCHOLOGY

THE RESURGENCE OF ELECTROCONVULSIVE THERAPY (ECT) ⊡

ECT is a psychiatric treatment in which electricity travels through the patient's brain to induce a seizure. In the 1930's a Hungarian psychiatrist, Ladislas von Meduna, noted a marked difference between the brains of epileptics and schizophrenics. He suggested that schizophrenics could be cured by artificially inducing an epileptic seizure. He injected camphor or a camphorlike compound into psychotic patients to bring about a seizure and reported miraculous improvements as a result of his treatment. Later it was discovered that seizures could be induced by applying electrodes to the patient's temples so that electricity would pass through the brain.

The use of ECT spread rapidly throughout Europe and the United States and eventually became abused by mental health workers. Although the procedure showed some efficacy for severe depression, mania, and schizophrena it was being used for many other forms of mental illness. Psychiatry then turned to drugs as ECT became less popular in the 1950's and 1960's. However, because the new drugs were not effective in treating severe mental illness in many patients and because the procedure has been markedly improved, ECT has experienced a resurgence.

IDEAS TO EXPLORE

- What are some short-term and long-term effects of ECT therapy?
- How is ECT different today from what it was when first developed?
- Why does the American Psychiatric Association and the National Institutes of Health support the use of ECT in specific cases of mental illness?
- Why do some organizations object to the use of ECT?
- What theories have been promulgated to explain the success of ECT in helping some patients?
- Has ECT been used on patients other than psychiatric with success?

SOURCES

The Sciences, Nov./Dec. 1989 p. 25.
R. Abrams

Electroconvulsive Therapy, R. Abrams,
Oxford University Press, 1988

SYMBOL KEY	
●	Topic of average difficulty
Δ	Long-term assignment
#	Project requires special facilities, equipment or supplies
o	Large public or college library required
*	Safety precautions required
+	Highly Technical; specialized knowledge required

MULTIPLE PERSONALITY DISORDER (MPD)

The Three Faces of Eve is a movie portraying the life of a person who has been said to have Multiple Personality Disorder or MPD. MPD is characterized by the "existence within the person of two or more distinct personalities, each of which is dominant at a particular time" according to the Diagnostic and Statistical Manual-IIIR (DSM-IIIR) of psychiatry. These other personalities are called "alters." A major cause of MPD is thought to be childhood abuse; developing multiple personalities becomes a defense against the pain and memory of the abuse. Although MPD is recognized as a separate illness in the DSM-IIIR, many psychiatrists feel that it is not real and some claim that patients can be suggested into behaving as though alters exist. While psychiatrists and other mental health care workers dispute the existence of the disorder, more and more patients have appeared in the mass media to talk about their lives.

IDEAS TO EXPLORE

- What are the symptoms of MPD? Give examples.
- What treatment strategies have been used?
- How common is the illness?
- What theories have been proposed to explain MPD?
- Can patients be cured of MPD?
- Why are some psychiatrists, psychologists, and psychotherapists doubtful about the existence of MPD?

SOURCES

Journal of Clinical Psychiatry
vol 47, 1986, p. 285.
F. W. Putnam et al.

Diagnostic and Statistical Manual, 3rd Edition Revised, American Psychiatric Association

SYMBOL KEY	
●	Topic of average difficulty
Δ	Long-term assignment
#	Project requires special facilities, equipment or supplies
○	Large public or college library required
∗	Safety precautions required
+	Highly Technical; specialized knowledge required

PSYCHOTHERAPY AND CANCER ● ○

A recent scientific study suggests that psychotherapy can extend the life of cancer patients. Patients with metastatic breast cancer were randomly assigned to therapy or control groups. All participants underwent radiation or chemotherapy. Those who received a year of group therapy lived an average of 18 months longer than those in the control group. Anxiety and pain were also reduced as a result of therapy.

IDEAS TO EXPLORE

- Are there other studies that show a positive effect of psychotherapy on cancer survival rates?
- If there is a positive effect, what are some possible explanations?

SOURCES

Lancet, Oct 14, 1989
D. Siegel
Stanford Medical School

SYMBOL KEY	
●	Topic of average difficulty
Δ	Long-term assignment
#	Project requires special facilities, equipment or supplies
○	Large public or college library required
*	Safety precautions required
+	Highly Technical; specialized knowledge required

THE EFFECT OF A RANDOM REWARD SYSTEM ON BEHAVIOR ⊡

PROJECT

Rewards can be distributed randomly, i.e., only some of the time after a task is completed, or consistently each time the task is completed. You can determine the effect of each reward system using hamsters or mice. Train two sets of animals to perform some task such as depressing a lever. For one group, reward each completion of the task with a food pellet. For the other group, distribute rewards randomly as group members depress the lever. To compare the efficacy of the two reward systems in shaping behavior, note the frequency with which the task is completed among the two groups. You could also remove the lever for a period of time and compare how long it takes to retrain members of the two groups to complete the task.

SOURCES

Any college introductory psychology text would contain material on reward systems. An example is
Psychology, 3rd Edition, 1981
C. Wortman, E. Loftus
Alfred A. Knopf Inc. New York

THE USE OF CORPORAL PUNISHMENT IN CHILD REARING ⊡

Fifty years ago the adage "Spare the rod and spoil the child" was considered unquestionable. However, today many parents, educators, psychologists, and pediatricians vehemently oppose the use of corporal punishment in raising children. Physical force is now thought to be counterproductive in shaping a child's behavior, serving to lessen self-worth and to teach that violence is an appropriate response to an undesirable situation. Despite the outcry amongst professionals to abstain from corporal punishment, a 1990 survey revealed that half of all parents strike or spank their children at least once a year. Also, 30 states permit physical punishment of students so that in the 1986–1987 academic year about one million children were spanked or beaten in schools. Between 10,000 and 20,000 of these children required medical attention. According to the National Center on Child Abuse Prevention, minority, poor, and handicapped students are much more likely to be the recipients of violence relative to their numbers. It should be noted that all continental European nations as well as Japan, Israel, Ireland, and Puerto Rico ban corporal punishment in public schools.

IDEAS TO EXPLORE

- What are the major schools of thought about the effectiveness of corporal punishment in child development today?
- Why do psychologists feel that corporal punishment is counterproductive and even harmful to the development of children?

PROJECT

Effect of Families on the Use of Corporal Punishment [•]

Do child-rearing practices depend more on the parent's generation or the particular practices of the child's grandparents? Conduct a survey of the frequency of spanking or hitting of children by members of both your parents' and grandparents' generation. Does the use of corporal punishment run in families or is it more dependent on the generation of the parents?

SOURCES

The National Center for the Study of Corporal Punishment and Alternatives in the Schools
833 Ritter Annex,
Temple University,
Philadelphia, PA 19122

National Committee for Prevention of Child Abuse
332 S. Michigan Avenue, Suite 1600
Chicago, IL 60604

12

SPACE SCIENCES

ASTEROIDS • ○

According to the "Big Bang" theory, asteroids are bodies with variable masses and velocities formed by the initial explosion that gave rise to the universe. Their past collisions with both the Earth and the Moon have produced craters as evidence of the impacts. Some scientists believe that it was such an impact that lead to the extinction of the dinosaurs. The smoke and debris caused a significant lowering of the temperature and decrease in the sunlight reaching the earth's surface. As a result, much vegetation and many plant communities perished.

Recent studies indicate that future collisions with asteroids are possible based on past evidence and suggest catastrophic effects unparalled in human history.

IDEAS TO EXPLORE

- What are asteroids?
- What are their chemical compositions?
- Are their orbits being observed?
- How probable is a collision between an asteroid and Earth? If an asteroid was on a collision course with the Earth, should attempts be made to alter its orbit or should it be destroyed?

PROJECT

Find the Force Needed to Deflect an Asteroid from the Earth's Path ○ +

Determine the equations derived from Newtonian mechanics that would be used to find the magnitude and direction of force needed to deflect an asteroid from the Earth's path. What variables must be known? Suggest a hypothetical scenario to demonstrate your approach. Perhaps you could use Basic or Fortran computer languages to program your calculations so that you can present the results of several possible scenarios.

SOURCES

NASA and the Jet Propulsion Laboratory in California are presently studying asteroids and their potential for collisions with Earth.

SYMBOL KEY	
●	Topic of average difficulty
Δ	Long-term assignment
#	Project requires special facilities, equipment or supplies
○	Large public or college library required
∗	Safety precautions required
+	Highly Technical; specialized knowledge required

OUR NEAREST NEIGHBORS ☐

Recent space expeditions by both manned and unmanned spacecraft are shedding light on some of our nearest planets in particular. On Feb 12, 1990, for example, the Galileo spacecraft photographed Venus from about 1.6 million kilometers away as it encircled the planet to accelerate towards a rendezvous with Jupiter in 1995. The resolution of the image is about 40 km. Information from less recent journeys by the Mariner 10 and Viking spacecraft are still being studied to understand mysteries surrounding Mercury and Mars. Meanwhile spectral measurements on earth continue to elucidate the atmospheric composition of the other planets of the solar system.

PROJECT

Description of a Planet ☐

Select one or more of the planets and describe what is presently known, data that has been obtained that has not be accounted for, and current theories about the origin and evolution of the planet.

SOURCES

Consult *Science*, *Nature*, and *Science News* for information on the planets.

Science News, Nov 4, 1989, p. 301.

Science News, Nov 11, 1989, p. 311.

Science News, June 16, 1990, p. 375.

THE VOYAGER SPACECRAFTS ⊡

Voyager spacecrafts in the thirteen years since their launching have contributed more to planetary science than all of the terrestrial observations made previously. Voyager 2 visited Jupiter, Saturn, (also visited by Voyager 1) Uranus, and Neptune. Voyager 1 is is expected to end its planned journey about 2015 when its plutonium source will be used up. By that time both Voyagers should have reached the heliopause (the edge of the solar system) and then drift through space.

IDEAS TO EXPLORE

- Trace the paths of the spacecrafts. A chart showing their journeys, along with dates of arrival would be helpful.
- Discuss some of the innovative features of the spacecrafts.
- How have the two missions added to our knowledge of the planets and their evolution?

SOURCES

E.C. Stone, Chief Project Scientist for the Voyager Mission
Physics Department, California Institute of Technology

NEPTUNE: SOME RECENT SURPRISES ⊡

Voyager 2 recently sent back images of the least known planet, Neptune. These photographs revealed giant storm systems and clouds never observed before on any of the planets. A hurricane the width of the earth appeared in the images. Prior to this newly acquired information Neptune was thought to be similar to its neighbor Uranus, a large sphere of molten rock and water surrounded by an atmosphere of hydrogen and helium mixed with methane. It is the methane that gives Neptune its aqua color.

Voyager also revealed six new moons whose compositions differ markedly from that of Neptune, implying that they were caught in the planet's gravitational field. Neptune's two other moons, Triton and Nereid, had been identified previously from terrestrial observation but little was known, especially about the largest moon, Triton. Voyager's high resolution images of Triton suggest a violent history. Large fissures and plumes of dark material along with a rippling terrain could have resulted from volcanic activity and melting. Triton's polar gap consists of methane and nitrogen ice at 37K, the coldest temperature measured so far in the solar system. Triton may have been an independent planet like Pluto before it was trapped by Neptune's gravitational field.

IDEAS TO EXPLORE

- How was Neptune originally discovered?
- What information had been obtained about the planet prior to the Voyager 2 mission?
- What are the current theories about the planet based on the new data?
- What questions remain?

SOURCES

Scientific American, Nov. 1989, p. 83.
J. Kinoshita

SYMBOL KEY	
●	Topic of average difficulty
Δ	Long-term assignment
#	Project requires special facilities, equipment or supplies
o	Large public or college library required
*	Safety precautions required
+	Highly Technical; specialized knowledge required

HEALTH EFFECTS OF SPACE TRAVEL ⊡

Adverse health effects are known to result from weightlessness (absence of gravity or microgravity) and radiation exposure. In addition, both physical and mental problems are associated with living in a small and limited environment for an extended period of time.

IDEAS TO EXPLORE

- What is the effect of weightlessness on the circulatory system? Experiments suggest, for example, a shrinkage of the heart muscle itself, perhaps a result of inactivity during space flight, especially in a gravity-free setting.
- What is the effect of weightlessness or microgravity on bones? Skylab astronauts showed a marked increase of blood calcium due to bone loss. If bone fractures occurred in space, how would weightlessness affect healing?
- Astronauts in a space shuttle 180,000 miles above the earth are bombarded by cosmic rays. Solar flares give rise to bursts of proton radiation. What may be the effects of this radiation?

PROJECT

Preventing Health Problems in Space [•]

A Mars mission is planned by 2010 and a permanent Mars base by 2025. Suggest health problems of astronauts travelling to Mars. Suggest and design possible prevention approaches.

SOURCES

Science, Sept./Oct. 1986, p. 47.
W. DeCampli

NASA, National Academy of Sciences, commission on major development in space science

SYMBOL KEY	
●	Topic of average difficulty
Δ	Long-term assignment
#	Project requires special facilities, equipment or supplies
o	Large public or college library required
*	Safety precautions required
+	Highly Technical; specialized knowledge required

VENUS ● ○

NASA (National Aeronautics and Space Administration) often suggests that the exploration of the planets enhances our understanding of the earth. Venus in particular is of interest because it is the most similar to the earth in primary properties although it differs significantly in secondary properties. It is thought that the evolution of Venus' atmosphere followed a very different path because of a collision between the earth and another body that led to the creation of our moon and our atmosphere. There are many questions about the origin, evolution, and geology of Venus, some of which will provide an indepth characterization of all global regions of Venus.

IDEAS TO EXPLORE

- What is known about the origin, evolution, and geology of Venus?
- What does its atmosphere consist of?
- What temperatures have been measured on the planet?
- Does Venus have mountain ranges? an ocean? volcanic activity?
- Describe the heat balance of the planet.
- What are some of the major mysteries about Venus?

PROJECT

Construct a Venus Globe ● ○

Much is known about the topography of Venus from orbiters and short-lived landers. You can use this data to construct a Venus globe using paper mache or some other suitable material. You might incorporate color to help show lowlands and rolling plains as well as mountainous regions.

SOURCES

Science, vol 247, p. 1191.
W. M. Kaula
Dept. of Earth and Space Sciences
University of California
Los Angeles, CA 90024

"Evolution of the Preplanetary Cloud and Formation of the Terrestrial Planets"
V. S. Safronov
NASA Publication 7771-55049
NASA, Washington, DC (1972)

Geology, vol 14, 1986, p. 14.
L. S. Grumplex et al

THE EVOLUTION OF CLIMATE ON PLANETS ⊡

A simple explanation for the fact that the earth's mean temperature is 15°C and therefore its surface contains liquid water necessary for life lies in Earth's distance from the sun. Venus is too hot because it is too close to the sun and Mars is frozen because it is too far away. Recent calculations reported by NASA (National Aeronautic and Space Administration), however, suggest another explanation for the significantly different climates of Venus, Earth, and Mars.

Venus, Earth, and Mars were once similar in many ways; the mineral composition of their surfaces and the gaseous components of their atmospheres were alike. Their now very different climates resulted from carbon dioxide cycling. The earth's cycling mechanism increases the concentration of carbon dioxide in the atmosphere when the surface cools and reduces it when the temperature rises. The mean temperature on Mars is −60°C because carbon dioxide stopped being cycled back into the atmosphere, while the temperature on Venus rose to 460°C because Venus could not remove carbon dioxide from the atmosphere. The NASA calculations attribute the earth's carbon dioxide cycle to primarily physical forces.

Carbon dioxide is removed from the atmosphere by rainwater which causes calcium-silicate minerals in rocks to erode. Thus calcium and bicarbonate ions are released into groundwater and eventually are incorporated into the calcium carbonate shells of marine life. As the organisms die, the shells settle into the sediment at the bottom of the oceans. Sediment spreads and slides under the ocean floor and is subjected to high temperatures and pressures in the earth's interior. There, calcium carbonate reacts with quartz to re-form silicate-containing rocks and in the process releases carbon dioxide, which eventually finds its way back to the atmosphere. (Another theory proposed by J. E. Lovelock and L. Margolis, called the Gaia hypothesis, attributes the earth's climate primarily to living organisms which removed carbon dioxide from the atmosphere.)

The NASA calculations indicate that Venus is too hot, not because it is too close to the sun, but because of too much carbon dioxide. Mars is frozen because it was too small to continue recycling carbon dioxide. If it were earth-size it would have been capable of supporting some form of life. The results of the NASA calculations suggest that the probability is good that habitable planets of other planetary systems exist.

IDEAS TO EXPLORE

- Why is earth habitable while other planets of our solar system are not?
- How do recent calculations explain the earth's climate? Why is Mars frozen? Venus too hot?
- How do the results of these calculations impact upon the probability of the existence of other habitable planets in other solar systems?

SOURCES

Kasting J. et al.
Scientific American, Sept. 1988, p. 90.

PART III

Resources

Resources for Research

Below are listed scientific societies, regional, national, and international organizations, and government agencies that may be helpful in researching your project or report. Also, a list of scientific supply companies is provided followed by a brief description of the national competition offered by the Westinghouse Electric Corporation for high school seniors and a brief description of the projects of some former winners.

Scientific Societies

AMERICAN ASSOCIATION FOR THE ADVANCEMENT OF SCIENCE
AMERICAN CHEMICAL SOCIETY
AMERICAN GEOPHYSICAL UNION
AMERICAN INSTITUTE OF BIOLOGICAL SCIENCES
AMERICAN INSTITUTE OF PHYSICS
AMERICAN PHYSICAL SOCIETY
AMERICAN PSYCHOLOGICAL ASSOCIATION
AMERICAN SOCIETY FOR AGRONOMY/
Crop Science Society of America/
Soil Science Society of America
AMERICAN SOCIETY FOR MICROBIOLOGY
AMERICAN SOCIETY OF PLANT PHYSIOLOGISTS
AMERICAN SOCIETY OF ZOOLOGISTS
AMERICAN VETERINARY MEDICAL ASSOCIATION
BIOPHYSICAL SOCIETY
ENGINEERING SOCIETY OF DETROIT
GEOLOGICAL SOCIETY OF AMERICA
INSTITUTE OF ELECTRICAL AND ELECTRONICS ENGINEERS

NATIONAL SOCIETY OF PROFESSIONAL ENGINEERS
OFFICE OF TECHNOLOGY ASSESSMENT
SOCIETY FOR RESEARCH IN CHILD DEVELOPMENT

The addresses of the above listed professional societies can be obtained from the

AMERICAN ASSOCIATION FOR THE ADVANCEMENT OF SCIENCE
1333 H Street N.W.
Washington, DC 20005

International, National, and Regional Organizations

International

WORLD HEALTH ORGANIZATION (WHO)
c/o UNITED NATIONS
405 East 42nd Street
New York, NY 10017
(212-963-6132)

UNITED NATIONS FOOD AND AGRICULTURE ORGANIZATION
U.N. Plaza
New York, NY 10017
(212-963-6039)

UNITED NATIONS ATOMIC ENERGY AGENCY
U.N. Plaza
New York, NY 10017
(212-963-6011)

UNITED NATIONS SCIENTIFIC & CULTURAL ORGANIZATION
U.N. Plaza
New York, NY 10017
(212-963-5995)

UNITED NATIONS ENVIRONMENTAL PROGRAMME
U.N. Plaza
New York, NY 10017
(212-963-8139)

National

CENTERS FOR DISEASE CONTROL
1600 Clifton Rd.
Atlanta, GA 30333
(404-639-3311)

SMITHSONIAN ASTROPHYSICAL OBSERVATORY (SAO)
60 Garden Street
Cambridge, MA 02138
(617-495-7461)

ENVIRONMENTAL DEFENSE FUND
257 Park Avenue South
New York, NY 10010
(212-505-2100)

CENTER FOR ENVIRONMENTAL PHYSIOLOGY
5632 Connecticut Avenue
P.O. Box 6359
Washington, DC 20015
(202-363-9575)

NATIONAL INSTITUTE OF HEALTH
or
NATIONAL CANCER INSTITUTE
9000 Rockville Pike
Bethesda, MD 20892-0001
(301-496-4000)

NATIONAL WILDLIFE FEDERATION
1400 Sixteenth Street N.W.
Washington, DC 20036-2266
(202-797-6800)

AGRICULTURE RESEARCH INSTITUTE (ARI)
9650 Rockville Pike
Bethesda, MD 20814
(301-530-7122)

Regional

NEW YORK ACADEMY OF SCIENCES
2 East 63rd Street
New York, NY 10021
(212-838-0230)

MEMORIAL SLOAN-KETTERING INSTITUTE
FOR CANCER RESEARCH
1275 York Avenue
New York, NY 10021
(1-800-422-6237)

NEW YORK BOTANICAL GARDEN
Southern Blvd. & 200th Street
Bronx, NY 10458
(212-220-8700)

WASHINGTON UNIVERSITY
REGIONAL PLANETARY IMAGE FACILITY
Department of Earth & Planetary Sciences
Box 1169
St. Louis, MO 63130
(314-889-5679)

McDONNELL CENTER FOR SPACE SCIENCES
Box 1105
St. Louis, MO 63130
(314-889-6225)

CARL HAYDEN BEE RESEARCH CENTER
2000 East Allen Road
Tucson, AZ 85719
(602-629-6380)

CEREAL RUST LABORATORY
UNIVERSITY OF MINNESOTA
St. Paul, MN 55108
(612-373-1300)

CLEMSON UNIVERSITY
PEE DEE RESEARCH & EDUCATION CENTER
FOR AGRICULTURE
P.O. Box 271
Florence, SC 29503
(803-662-3526)

PURDUE UNIVERSITY
SOUTHERN INDIANA-PURDUE AGRICULTURE
CENTER
R.R.2
Dubois, IN 47527
(812-678-3401)

GEORGIA CROP REPORTING SERVICE
Federal Building, Suite 320
Athens, GA 30613
(404-546-2236)

NEW ENGLAND PLANT, SOIL & WATER LABORATORY (NEPSWL)
UNIVERSITY OF MAINE AT ORONO
Orono, ME 04469
(207-581-2216)

NEBRASKA STATEWIDE ARBORETUM (NSA)
112 Forestry Sciences Laboratory
Lincoln, NE 68583
(402-472-2971)

AMERICAN HEALTH FOUNDATION
320 East 42 Street
New York, NY 10017
(212-953-1900)

BAYLOR COLLEGE OF MEDICINE
BIRTH DEFECTS CENTER
6621 Fannen Street
Houston, TX 77030
(713-791-3261)

CENTER OF ALLERGY & IMMUNOLOGICAL DISORDERS
1200 Moursund Avenue
Houston, TX 77030
(713-791-4219)

BOYCE THOMPSONS INSTITUTE FOR PLANT RESEARCH
CORNELL UNIVERSITY
Tower Road
Ithaca, NY 14853
(607-257-2030)

BROWN UNIVERSITY
ANIMAL CARE FACILITY
Box C
Providence, RI 02912
(401-863-3223)

CASE WESTERN RESERVE UNIVERSITY
DEVELOPMENTAL BIOLOGY CENTER (DBC)
Cleveland, OH 44106
(216-368-3430)

COLUMBIA UNIVERSITY
CORNELL PLANTATIONS
1 Plantation Rd.
Ithaca, NY 14850
(607-256-3020)

DUKE UNIVERSITY
MARINE BIOMEDICAL CENTER (MBC)
Duke University
Beaufort, NC 28516
(919-728-2111)

HARVARD UNIVERSITY
KRESGE CENTER FOR ENVIRONMENTAL HEALTH
665 Huntington Avenue
Boston, MA 62115
(617-732-1272)

CENTER FOR SCIENCE & TECHNOLOGY
RENSSELAER POLYTECHNIC INSTITUTE
725 Park Avenue
New York, NY 10021
(212-772-8120)

NATIONAL AERONAUTICS & SPACE ADMINISTRATION (NASA)
GODARD SPACE FLIGHT CENTER
2880 Broadway
New York, NY 10025
(212-678-5500)

THE CENTER FOR SPORTS MEDICINE
445 5th Avenue
New York, NY 10016
(212-685-9633)

CENTER FOR THE STUDY OF ANOREXIA
1 West 91st Street
New York, NY 10024
(212-595-3449)

CENTER FOR BULIMIA & RELATED DISEASES
8 Gramercy Park South
New York, NY
(212-254-2809)

Government Agencies

UNITED STATES ENVIRONMENTAL PROTECTION AGENCY(USEPA)
401 M Street S.W.
Washington, DC 20460
(202-260-2090)

UNITED STATES DEPARTMENT OF AGRICULTURE (USDA)
ARS VISITORS CENTER
Bldg. 302, BARC-East
Powdermill, Rd.
Beltsville, MD 20705
(301-344-2403)

DEPARTMENT OF JUSTICE
LAND & NATURAL RESOURCES DIVISION
200 Mamaroneck Avenue
White Plains, NY
(914-949-4806)

DEPARTMENT OF JUSTICE
DRUG ENFORCEMENT ADMINISTRATION
600-700 Army Navy Drive
Arlington, VA 20202
(202-307-1000)

FEDERAL MARITIME COMMISSION
6 World Trade Center
New York, NY 10047
(212-264-1425)

NATIONAL OCEANIC & ATMOSPHERIC ADMINISTRATION
NATIONAL WEATHER SERVICE
Rockefeller Plaza
New York, NY 10020
(212-315-2705)

UNITED STATES FISH AND WILDLIFE
1849 C Street N.W.
Washington, DC 20240
(202-208-5634)

Scientific Supply Companies

BAXTER SCIENTIFIC PRODUCTS
1430 Waukegan Rd
McGraw Park, IL 60085-6787

SARGENT-WELCH SCIENTIFIC COMPANY
(Biology, Chemistry, Physics, Earth/Space, Environmental Sciences)
7300 North Linden Avenue
P.O. Box 1026
Skokie, IL 60077

THOMAS SCIENTIFIC APPARATUS AND REAGENTS
99 High Hill Road at I-295
P.O. Box 99
Swedesboro, NJ 08085-0099

FISHER SCIENTIFIC
711 Forbes Avenue
Pittsburgh, PA 15219

(Sold by Fisher:
Bay Water Technology Manual
Standard Methods Manual for Water Analysis
EPA approved analytical methods)

WESTINGHOUSE SCIENCE TALENT SEARCH

Since 1942 the Westinghouse Electric Corporation has offered scholarships to high school seniors who are semifinalists and finalists in their annual, nationwide science competition. About 1,500 high school seniors participate every year. The entry consists of a written report describing the student's research along with a completed entry form. Eight distinguished scientists from different disciplines serve as the judges. Semifinalists and finalists not only receive scholarships but also are recommended to colleges and universities for admission and financial assistance. Finalists also visit Washington, DC where they meet leading scientists and their Congressional representatives. They are subjects of frequent media interviews and past winners have met with the President. Ten winning projects are briefly described below.

- Experiments suggesting the biochemical pathway involving cholesterol in the initiation of atherosclerosis (Mitchell Wong, New York)
- The effects of ultraviolet-B radiation on the mortality and reproductive capacity of Daphnia pulex, a water flea important to the health of freshwater ponds and lakes (Renee Lynette Doney, Montana)
- The effects of stress on problem solving (Jason W. Victor, New York)
- Determination of the dimension of fractals generated by Pascal's Triangle (Ashley Melia Reiter, North Carolina)
- Investigations on cell mobility elucidating the role of ATP and calcium ions on ciliary movement (Johannes Sebastian Schlondorff, New York)
- The effects of copper, manganese, and zinc ions on the production of betacarotene by Romaine lettuce (Susan Elaine Criss, Pennsylvania)

- The relationship between friendships and morale among residents of nursing homes (Petal Pearl Haynes, New York)
- Investigation of factors affecting the stability of tautomers (Daniel Moshe Skovronsky, Virginia)
- Relationship between musical talent and math aptitude in high school students (Tara Sophia Bahna-James, New York)
- The effect of stress on eating habits (Summee Louise Kim, New York)

The Science Talent Search is administered by Science Service, a nonprofit institution. For information write:

SCIENCE SERVICE
1714 N Street N.W.
Washington, DC 20036
(202)785-2255

ALSO FROM 

College Entrance

Preparation for the PSAT/NMSQT
Preparation for the SAT - Scholastic Aptitude Test
Verbal Workbook for the SAT
Mathematics Workbook for the SAT
ACT: American College Testing Program
Mathematics Workbook for the New ACT
English Workbook for the New ACT
College Board Achievement Test in American History and Social Studies
College Board Achievement Test in Biology
College Board Achievement Test in Chemistry
College Board Achievement Test in English Composition
College Board Achievement Test in Mathematics: Level I
College Board Achievement Test in Mathematics: Level II
College Board Achievement Test in Spanish
Advanced Placement Examination in American History
Advanced Placement Examination in Mathematics
Advanced Placement Examination in English Composition and Literature
Advanced Placement Examination in Computer Science
Advanced Placement Examination in Biology
Advanced Placement Examination in Chemistry
College English Placement and Proficiency Exams
Preparation for the CLEP: College-Level Examination Program
SuperCourse for College Board Achievement Test
SuperCourse for the SAT
SuperCourse for the ACT
SuperCourse for the TOEFL
TOEFL: Test of English as a Foreign Language
TOEFL Skills for Top Scores
TOEFL Reading Comprehension and Vocabulary Workbook
TOEFL Grammar Workbook
Cram Course for the SAT
Cram Course for the ACT

PROJECT NOTES

Project Notes

Project Notes

Project Notes